Self-Sufficiency Cheese Making

Self-Sufficiency

Cheese Making

Rita Ash

First published in 2009 by New Holland Publishers (UK) Ltd
London • Cape Town • Sydney • Auckland

Garfield House
86–88 Edgware Rd
London W2 2EA
United Kingdom

80 McKenzie Street
Cape Town 01
South Africa

Unit 1
66 Gibbes Street
Chatswood
NSW 2067
Australia

218 Lake Road
Northcote
Auckland
New Zealand

ISBN 978 184773 461 7

Senior Editor: Emma Pattison
Designer: Melissa Hobbs at e-Digital Design
Main illustrations: Michael Stones
All other artwork: e-Digital Design
Production: Laurence Poos
Editorial Direction: Rosemary Wilkinson

1 3 5 7 9 10 8 6 4 2

Reproduction by Pica Digital PTE Ltd, Singapore
Printed and bound by C&C Offset Printing Co Ltd, China

CONTENTS

INTRODUCTION

I learnt to make cheese with the aid of a 1947 British Ministry of Food pamphlet found in the library of an agricultural college and by asking questions of cheese-makers. For over 30 years I have been making cheese, teaching others to make cheese and facilitating small-scale commercial cheese-making enterprises. I have made cheese from many types of milk and in many different environments, and I can vouch that good cheese can be made under the most primitive conditions, even in a mud hut on a charcoal burner.

This book, not intended for those living in a mud hut but for those with access to modern facilities, explains the processes involved in making cheese, covers which cheeses you might like to make and includes a selection of recipes for both soft and hard cheeses, with some suggestions for their usage. I've also given some advice for expanding from making cheese in the kitchen to small-scale commercial production.

Traditionally the skill of cheese making has been passed by word of mouth as relatively few formal instructions for making cheese existed. Despite this lack of direction cheese makers the world over have produced delicious cheeses which were the forerunners of today's classic cheeses.

You do not need to be scientifically trained to make cheese, except, perhaps, when vast quantities are at risk. Nor do you need to be an accomplished cook, although being a good judge of taste, texture and smell helps a lot. Cheese making really is a simple process and I hope that with the help of this book, and by simply giving it a go, you will soon be able to make delicious cheeses to enjoy at home and perhaps sell too.

Throughout the text the I have tried to pass on not only my accumulated knowledge and experience of cheese making, but also the sense of pleasure and satisfaction I still get from having my hands in a vat of warm milk and caring for the cheeses produced.

Rita Ash

A brief history

Legend has it that the first cheese was made by a herdsman quite by accident, and there is archeological evidence of the existence of cheese dating back to 5,000 BC. While the exact origin of cheese making is unclear, it seems likely that cheese was first made in ancient times to preserve the milk of lactating animals rather than wasting the spare milk.

How it began

Legend attributes the first cheese to an accident. A roving herdsman, perhaps a shepherd or a goatherder, had one of his animals give birth to a lamb or kid which shortly afterwards died leaving it's mother in milk. The herdsman took the stomach from the dead youngster, rinsed it in a stream, tied up the exit hole to form a bag and then milked out the lactating mother and stored the milk in the stomach bag for later consumption. Hanging the bag from his pack, he travelled on. Later, feeling thirsty, he went to the bag intending to drink the milk, only to find he had a lump of white solid material floating in clear, greenish water. He drank the liquid, which had a pleasant flavour, and ate the solid which satisfied his hunger. This is how, it is said, the first cheese was formed.

Much more likely, but without the legend, is that the cheese had formed wherever lactating animals had been milked into any sort of container rather than drunk straight from the teat. Any milk left overnight, intentionally or otherwise, would have made cheese by the morning and would have been consumed as a matter of course. If milk, or water for that matter, had to be transported, the stomach from a dead animal would have been used as a ready-made bag. However, it seems unlikely that this knowledge and activity led to any deliberate production of cheese, rather that it was an alternative to wasting spare milk.

Whatever is the fact of the matter, there is no doubt that milk left in a stomach bag would have gone sour in the warmth, and had the stomach been that of a suckling lamb or kid, within a few hours it would have formed a solid due to the residual rennin in the stomach lining. The result would have been curds and whey.

The milking and herding of animals

As well as being a source of milk, animals also, of course, provided meat. Male kids and ram lambs would have been slaughtered along with animals no longer of economic use. None of the edible meat would have been wasted. That not eaten would have been cut or torn into strips and dried for later consumption. Amongst the preserved meat would have been the cleaned stomach of the slaughtered animal. Both ingredients for the making of cheese, being milk and rennin, were therefore readily available but two vital constituents were missing: the understanding of the necessity of the stomach or part of the stomach of a suckling youngster in the cheese making process and a container in which to hold the milk from a number of animals, enough to make it a worthwhile exercise. The combining of stomach meat and sour milk at meal-times probably provided the first connection but the second solution would come much later with the advent of tools to work wood and the skills to make pottery.

There are no indisputable records of the origins of cheese but there is certainly archeological evidence of the milking and herding of animals and the existence of cheese and cheese making utensils dating from about 5,000 BC. These include Libyan cave paintings and an Assyrian relief showing cattle being driven by a cowherd carrying what are believed to be bags of milk. The remains of a substance which scientists are certain to have once been cheese have been found in an Egyptian tomb dating from 3,000 to 2,000 BC.

Early cheese making

It seems most likely that cheese 'happened' and was developed into a foodstuff wherever there were communities, or that nomadic husbandmen, already experienced in the practice of cheese making and looking for pastures new for their stock, spread their knowledge widely. Certainly by the times of the Greek and Roman empires, cheese formed an established part of the daily diet.

Most primitive cheese would have been a simple lactic cheese, made by allowing milk to sour and putting the resulting soft curd into cloths or fine baskets to drain. The nomadic Bedouin in Jordan still make their cheese in exactly this manner, as do cheese makers in parts of India, but the method is slow and the yield is low. Traditionally, a setting agent is not used when making cheese using this method.

Although the most common form of coagulant was, and still is, the enzyme rennin, traditionally obtained by soaking a strip of suckler's stomach lining in the cheese milk, many other agents have been tried as coagulants for use in areas where animals were not part of the diet. Plant and fruit juices, flowers and bark have all been tried with differing success.

Cheese production

By the time of the Roman Empire the making of cheese was not far different from that which we are familiar with today and had become widespread and varied. Each country and each area of that country had their preferred methods of production according to their circumstances, and these were passed from generation to generation by example and word of mouth. The passage of Roman legions through these countries, together with the indigenous camp followers, would have influenced 'home' recipes and methods and introduced different skills and equipment. The Vikings brought Scandinavian ideas and introduced their cheese making skills to many regions and with the spread of Christianity monks came into being who set up their own farms and dairies making cheese, some of which, like Wensleydale, are still around today.

The monks were also scholars and could therefore write, and it was about this time that recipes and methods began to be recorded in the monastery archives. These writings were, however, not available to the farmhouse cheese maker who still relied on word of mouth and watching others.

As the demand for new varieties of cheese increased, so trade spread between cities and countries and cheese markets were established wherever trading took place. Cheese was still produced on a fairly small scale in farm dairies, but once the industrial revolution gained pace, cheese making grew into an industry. In 1838 Justus von Liebig, a German chemist, explained in scientific terms the fermentation (or ripening) processes of cheese making, a phenomenon known but not understood for millennia.

All was not, however, plain sailing. The middle of the nineteenth century saw outbreaks of cattle plague which decimated the town and city herds and forced the manufacturers and processors to look further afield for their milk. Fortunately for them, by this time the railways now reached throughout the countryside and were able to bring the milk to the centres quickly and without spoilage.

Pasteurization

Forty years after Liebig Louis Pasteur published his findings on lactic fermentations, recommending that milk for cheese making be pre-heated to destroy harmful bacteria. Pasteurization and heat treatment became widespread and essential in the large cheese making factories where losses through spoilage of milk or product could have been disastrous.

Pasteurization was not, however, the practice in farmhouse cheese making where milk was generally produced on the premises and less than a day old when made into cheese. With a twice daily milking regime the evening milk was allowed to stand overnight before the cream was skimmed off and the skimmed milk mixed with the morning milk to provide a starter to the ripening process.

This process was superseded in the 1930's by the introduction of pure 'mother cultures' which could be bought from dairy suppliers. The liquid culture came in a sealed bottle and was propagated overnight before being added to the cheese milk in the morning. These cultures were in use well into the 1970's until the introduction of packeted, freeze-dried starters. All of these pure starters were developed to reduce the risk of contamination during cheese making by ensuring that predominantly pure and safe bacteria were present in the milk at the start of the process. They also guarantee a continuity of flavour and quality when all other facets of the production are constant.

The cheese factory

The cheese factory, albeit on a smaller scale than our big creameries of today, came into being around 1860. At this time many of the small scale dairies ceased to make their own cheese and instead sold their milk to the factories. Nevertheless there remained a hard core of small producers, selling to a local market, some of whom made good cheese – others not so good. A contemporary rhyme regarding Essex cheese states:

They that made me were uncivil
For they made me harder than the Devil
Knives won't cut me, fire won't sweat me
Dogs bark at me but won't eat me.

Samuel Pepys recalls in his diary returning home one night to find his wife very angry "at her people for grumbling to eat Suffolk cheese". However, much regional cheese was of a quality which was average–good judged by the standards of the time, and survived in ever-decreasing quantities until the Second World War when cheese production was strictly regulated. At this time many small production cheeses were lost, Dorset Blue Vinney among them, as it was hard cheese, not regional specialities, which was in demand for civilian rations and the services.

There was a revival of interest in our supposedly forgotten cheeses in the early 1980s when a letter was published in the Western Morning News, a British regional newspaper, asking for remembered old cheese recipes. Among the replies were recipes for a Cornish cheese, a Cambridge cheese, a small-scale Cheddar and a delightful poem (see page 16) attributed to a Doctor Jenner on the making of a soft cheese. There is no corroborative evidence that this was the Doctor Jenner of smallpox fame but as he was born (in 1749) and worked in the vale of Berkeley, which is another name for, and the home of, Gloucester cheese, it seems very likely.

To make a Soft Cheese by the late Dr. Jenner

Would you make a soft cheese? Then I'll tell you how.
Take a gallon of milk, quite fresh from the cow,
Ere the rennet is added and the dairyman's daughter
Must throw in a quart of the cleanest spring water.
When perfectly curdled, so white, so nice,
You must take it all out of the dish with a slice
And put it 'thout breaking with care in the vat
With a cheesecloth in the bottom – be sure to mind that.
This delicate matter take care not to squeeze
But fill as the whey passes off by degrees.
Next day you may turn it and do not be loth
To wipe it quite dry with a fine linen cloth.
That this must be done you cannot well doubt
As long as you can see any whey oozing out.
The cheese is now finished and nice it will be
If enveloped in leaves from the green ashen tree,
Or what will do better, at least full as well,
In nettles just pluck'd from the banks of the dell.

As people began to travel more so they developed a taste for the range of cheeses available in other countries and large quantities of cheese began to be exported and imported. To encourage home production, diversification grants were made available to farmers and smallholders and 'on farm' cheese making was encouraged. The wheel had turned full circle. Agricultural colleges, rural training boards and the milk marketing agencies all ran courses in cheese making at low cost for interested parties, and associations of smallholders promoted cheese making courses along with other skills.

At this time it was usual for dairy farmers to keep cows and process a little of the milk into cheese. However, while smallholders may have had a suckler cow, most kept goats, and in the late 1970's a revival of sheep milking occurred, both of which led to a substantial increase in the variety of cheese being made outside of the factories. The willingness of retailers to market these smallholder-made cheeses led to increases in the size of goat herds and sheep-milking enterprises, and the opening of delicatessens, farmer's and country markets widened the opportunities for marketing the 'farm' cheese.

Today there is a wide range of farm or smallholder cheese making enterprises, but it is likely that all will have had to go through a long and restrictive process before they could market their goods. Most countries have legislation concerned with protecting the consumer from poor practice in the dairy, often requiring that every premise preparing food for human consumption follows a set of registrations, inspections and certifications including writing a hazard analysis.

For the producer this is likely to mean that the whole process, including all cleaning down and hygiene methods, must be documented together with a program for dealing with any failures in the process. Visits from environmental health officers must be expected, during which time the premises and records will be checked. However, in some countries it is still not illegal to make cheese from milk which has not been heat treated or pasteurized, but it must usually comply with labelling regulations.

Cheese making remains of wide interest, confirmed by the variety of cheeses available. With a pedigree stretching back millennia and spanning continents, how could it be otherwise?

The Basics

Before you can begin the fun of making cheese yourself it is useful to understand the principles of cheese making. A sound knowledge of the science involved should help you to prevent errors and a basic grasp of the economics of cheese making will be invaluable should you choose to produce cheese on a larger scale.

What is cheese?

Cheese is basically sour milk with most of the water removed by careful manipulation. If you leave any milk out of a refrigerator for too long it will eventually 'go sour' and become the simplest form of cheese. The natural bacteria in milk, lactobacillus, reacts with the lactose in the milk and releases lactic acid. The lactic acid will coagulate, curdle or set some of the protein in the milk, making it insoluble in water and causing the water to separate from the solids. This fluid, containing the non-coagulated protein, some lactose, vitamins and minerals, is called whey.

When a young animal suckles milk from its mother, the milk, if it remained in liquid form, would pass far too easily through the animal's digestive tract to allow for proper absorption and nutrition, so nature has provided a natural acid in the stomach to coagulate the milk. To assist the solidification of the protein the stomach contains an enzyme, rennin, which also has the ability to coagulate proteins. As a result the milk in the young animal's system becomes semi-solid and passes more slowly through the intestine, allowing the nourishment needed for growth and development to be absorbed. The excess water is passed from the body separately.

All cheeses, whether soft, semi-hard or hard, are basically copying this natural process in a controlled and hygienic way. The separation of curd (the solidified proteins) from the whey can be achieved by adding an acid such as lemon juice or vinegar to fresh milk, allowing it to set, then drawing off the whey using cloths or baskets.

Traditionally, farm-made cheese was acidified by leaving the milk produced from the evening milking out overnight so that by morning it would be acidic from the action of the natural bacteria. The fresh morning milk was then added to this 'sour' milk, the combination of which gave a suitable acidity to begin cheese-making. The problem with this method is that undesirable bacteria, either in the milk or from the atmosphere, could also be multiplying. Some of these could be pathogens, which are organisms that cause illnesses in humans, or there could be organisms present that produce gas or unwanted flavours in the cheese.

The milk we use to make cheese today is usually heat treated to destroy this bacteria and is stored and handled in such a way as to avoid atmospheric contamination. Pure, laboratory selected bacteria known as starter are added to replace the natural bacteria. Even raw or untreated milk has a proportion of starter added so that the speed of multiplication is weighted in favour of the selected, rather than the random, bacteria.

In most cases a quicker and firmer coagulation is required. For these cheeses, at a point made clear in the recipe, an enzyme will be added to the milk that replicates the coagulation in the suckling animals stomach. Rennet replaces the natural rennin and is prepared from calf stomachs or, more frequently, from a micro-organism.

Cheese is a product of necessity. Milk is a very important source of human nutrition, but it is prone to spoilage, difficult to transport and has a limited shelf life. Nomadic tribes could only use milk from the animals that they had with them in liquid form, but could easily transport it once it had been processed into cheese. Farmers and smallholders throughout the ages and in most of the world have converted excess milk at peak lactation into cheese for use when the milk flow dropped off. In the days before refrigeration and freezing, milk was converted into cheese daily in spring and summer for use in the winter months. Travellers and day workers carried bread and cheese for their food, hence the emergence of the Ploughman's Lunch, and history tells us that during the Roman occupation of Britain all Cheshire cheese was taken for use by the Roman Army on their travels.

Note

Throughout the book, mold refers to the former for the curd, and mould to the micro-organisms which cause coating and veining of the cheese.

Types of cheese

Soft, un-molded cheeses include acid curd, which is simply set soured milk drained through cloth or basket; cream cheese, which is made in the same way but with added cream; and cottage cheese, which is filled into pots or boxes after draining and seasoning.

Soft molded cheeses are low rennet, low acid, curd cheeses that are packed into various shapes of mold for drainage. Soft molded cheeses such as Camembert and Brie are encouraged, by the addition of a special culture, to grow a white mould coat and soft blue veined cheeses have a different culture added to create the blue veins. These cheeses have a limited shelf life and need constant refrigeration.

Semi-hard cheeses are made using starter and rennet. Frequently they are drained through a sieve or cloth and put into the mold with little or no added pressure, leaving the curd to drain under its own weight. Amongst these are some of the blue veined cheeses, whereby the lack of pressure allows a looser packed cheese in which the veins can grow.

Hard cheeses are also made with starter and rennet. The curd is worked to expel the whey and the cheese is packed into the mold with pressure applied to consolidate the pieces of curd into a smooth whole with no voids. These cheeses tend to be the ones that take the longest to mature and can be kept for a considerable length of time without refrigeration.

The development of regional cheeses

Originally, the variety of cheese was specific to the household in which it was made. The method was passed down by word of mouth from one generation to the next. Later, groups of villages, growing to counties and regions, developed cheeses particular to their areas. The method was written down so that farm wives and dairymaids worked to the same instructions. Later still, travelling dairy instructors went from village-to-village holding training courses at selected farms, at which the cheese makers of the district were taught proper hygiene and handling of milk as well as the 'correct' way to make their cheese.

Regional variations

Regional cheeses developed over time rather than being specifically designed. They had to fit the volume and type of milk available, but above all else the stages of production had to fit around the cheese maker's other commitments.

Cheddar cheese – which is probably the most labour intensive cheese – would have taken most of a working day to attend to. Yet because this cheese originated from the large wealthy dairies in the west of England, farmers could afford to employ dairy maids whose sole job was to tend to the dairy; their whole day was dedicated to making the butter, clotting the cream and making and caring for the cheese.

Caerphilly cheese is a simpler cheese to make as in this Welsh region farms were smaller, less wealthy and much more difficult to work, and so not as much time could be dedicated to cheese making. More often than not the farmers wives made Caerphilly along with all their other duties and it therefore needed to be quicker and less labour intensive. It is said that Caerphilly was sometimes made in a communal dairy. The farmer would take his milk to the dairy for processing and when the curd was ready a bell would be rung. Villagers would then run to the dairy to help with the cutting and packing, which had to be done before the curd got cold.

Lancashire cheese is made over 2 or even 3 days. Lancashire farmers often kept only a few house cows which meant there was insufficient milk on a daily basis to make a reasonable sized cheese, so the daily production was made into curd and set aside. The second day's curd was added to the first and then the third day's when it was ready to make up the appropriate quantity before packing for pressing.

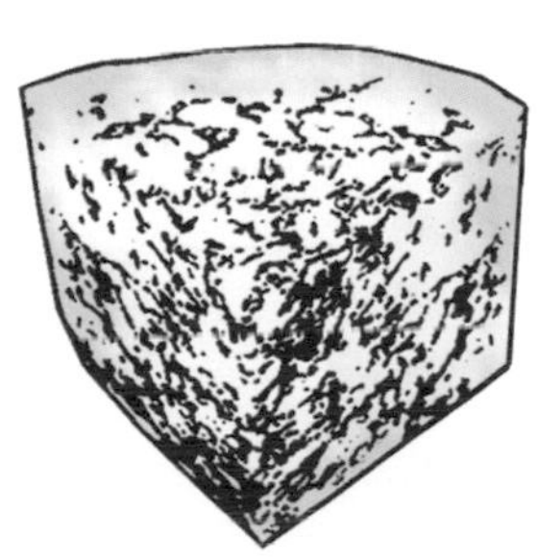

Stilton cheese takes many hours to make in the old farmhouse way. The milk is started and renneted and, when set, it is transferred to containers lined with cloths. These are knotted and left for the whey to collect around them. The whey is removed at irregular intervals throughout the day enabling the cheese maker to get on with other farming tasks. At the end of the afternoon's work, the cheese could be put into the mold. The blue veining would originally have been accidental, caused by mould growing in the cracks of the loosely packed and unpressed curd.

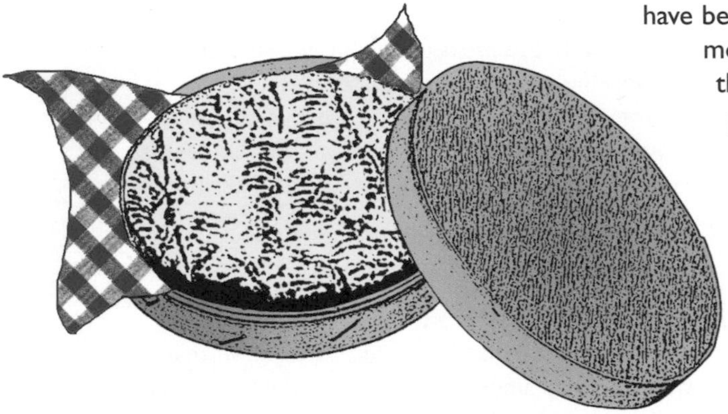

Camembert cheese is said to have originated in France in 1791 and is named after the hamlet in which it was first produced. It is similar in production to Brie but there is a difference in the flavour of a farmhouse Camembert partly due to the size and ripening time. Brie is produced in several areas of France, but as neither Brie nor Camembert are protected names, they are also made in many other parts of the world.

Parmesan, originally from Parma in Italy, is also known by a number of other names. It is a skim milk cheese that is heavily brined after molding and pressing. The cheese is matured for many months (14 months minimum in the USA) and will keep indefinitely despite becoming very hard. Many other cheeses are made from hand skimmed milk, which is the milk that remains after it has stood for some time and the cream has been taken from the top for clotting or butter making. In most cases these cheeses were originally for home use and not for general sale, although the market for low fat cheese has grown in recent years.

The cheeses mentioned so far are all traditionally cow's milk cheeses, but goat, sheep and buffalo milk cheeses have also been made throughout the world for centuries. Many of these are versions of traditional cow's milk cheeses, and those that are historically made from other milk are now copied in the more readily available cows milk.

Mozzarella originated in the Italian provinces of Caserta and Salernois. While it is traditionally made from buffalo milk obtained from a single early morning milking, most commercial Mozzarella is now made from cow's milk.

Wensleydale, now also primarily made from cows' milk, was originally made from sheep milk by the monks who introduced it to Britain. Goats' milk cheeses frequently use similar or the same production processes as cows' milk cheeses. Specific goats' milk cheeses, such as Chevre or Chevretin, are still produced, but cheese makers who use goats' milk tend to develop their own recipes.

Why make cheese?

The short answer to the question 'Why make cheese?' is the enormous amount of pleasure you will get from making it and the pride you will feel in putting your own hand-made cheese on the table – be it a little white herb-smelling soft cheese on a plate, cubes of smooth Feta in a salad, or the golden mound of a hard cheese with its smooth coat and appetising smell on the cheese board. For the more adventurous, enjoy the delicate blue veining of your mould-ripened cheese, or the white felt coat surrounding the melting centre of a soft cheese. All this enjoyment should more than make up for the time and effort spent making and caring for your cheese.

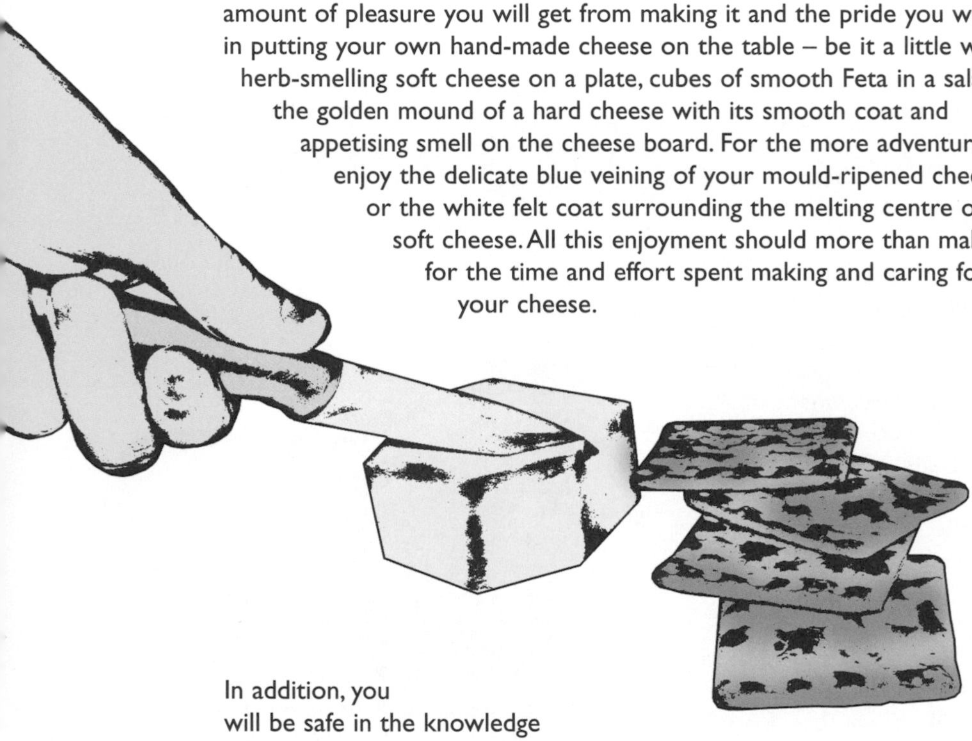

In addition, you will be safe in the knowledge that you have the skills to turn quantities of wet milk into something quite different, which most people have to shop for. You are also part of the growing number of people keeping alive this ancient and worthwhile craft. Simply choose the cheese appropriate to the time you have whilst fitting it around your normal life, put all your equipment near to hand and get ready to enjoy yourself!

The waiting game

There are lots of waiting times in cheese making; when the milk is ripening (souring), when the rennet is setting the curd and when the curd is draining. But during these times you will be able to do other small tasks around the house, or even read a good book. Stirring the milk and the curd with your hands is also quite time consuming but very satisfying. Don't be tempted to try and speed up the process by using a spoon as hands are more gentle on the curd than tools and smelling the milk on the back of your hand is a good guide to the ripeness of the milk after starter has been added. Besides there is real pleasure in hands gently stirring a pot of warm milk and I would go so far as to say that it's therapeutic.

Having hard cheeses in store for the weeks that they are maturing demands a few minutes each day to turn and clean them, but you will get quite fond of your cheeses and impatient to try them. Resist this as immature cheese is a disappointment and one more week can make all the difference. Once you have made a cheese that you are pleased with, cheese making will not be a chore. You will look back with pride, and forward to the next time with anticipation.

Personalizing your cheese

Most cheese makers start by making simple soft cheeses and plain hard or semi-hard cheeses, taking instruction from written recipes, but successful cheese is that which comes easiest to the cheese maker. You may well find that you create a personalized cheese that is a simple, convenient variation on a known recipe. In any case, hand-made cheese often varies from batch to batch. A change in the milk, a slight temperature rise or drop, a little less or more rennet, a firmer set, a cleaner cut or a higher scald temperature will all alter the finished cheese to a greater or lesser extent. If you want to personalize your cheese do not be afraid to make adjustments, but do bear in mind that the basic processing stages must be followed.

Additives and flavourings such as herbs, garlic, mustard, onion, beer, wine, cider and fruit can all been used successfully either in the processing or the finishing of cheese (see page 86). Don't be afraid to experiment!

Cheese economics

The economics of making your own cheese depends on where your milk comes from and what type of cheese you are making. It takes approximately 10 litres (17½ pints) of milk to make a kilogram (2 lb 4 oz) of hard cheese and you can make a 10 cm (4 in) diameter, 3 cm (1¼ in) deep round soft cheese from a litre (1¾ pints) of milk.

If you are lucky enough to have a surplus of your own milk from cows, sheep or goats then the economics are obvious – better to make cheese than waste good milk. But if you are buying the milk then the price you pay will dictate the added value. Some smallholders and goat keepers will have excess milk at high yielding periods that can be bought or traded. Local dairies may be persuaded to sell you small quantities of milk but it has, in recent years, become more difficult for dairy farmers in many countries to sell raw milk since contracts, legislation and sealed tanks were introduced. A small producer may have limited amounts of milk daily that can be collected up to make the quantity you need to work with. All fresh milk for cheese making can be frozen in blocks in an efficient domestic freezer, or kept refrigerated for up to 5 days at 4° C (39° F) or below.

If you are buying milk at retail price there is still a saving to be made by making your own cheese as the milk needed to make a quantity of cheese will almost always be cheaper than buying that quantity of cheese ready-made. When you add together the pleasure in the making of the cheese, the pride in the product, the infinite variations you can produce and the financial saving, it's easy to see why you should have a go at cheese making.

Making cheese

Here is everything you'll need to know to start making cheese, from the raw ingredients and equipment needed to organizing your working and storage areas. The basic method included in this chapter forms the basis for all of the recipes in this book and should be read carefully before attempting to make your first cheese.

Milk

In theory the milk of any lactating animal can be used to make cheese. In practice the milk of cows, goats, sheep and buffalo are most commonly used, although a form of curd can be prepared from the milk of reindeer and camel and also historically from asses and mares.

The majority of cheese made today is commercially produced from the mixed milk of a number of cows. The milk is often standardized, that is to say that the butterfat content has been adjusted by separating off excess cream. The milk will be pasteurized by heating rapidly to 72° C (161° F) and then immediately cooled. It will also have been homogenized, which breaks down the fat globules and mixes them thoroughly with the solids so that the fat does not rise to the top of the milk when left to stand.

Butterfat

Cows' milk used for home produced cheese may need some adjustment to the butterfat content. Too much butterfat is not good for most cheese making – it gives a loose, greasy curd and also excess butterfat can be lost to the whey, which is wasteful. If using a high butterfat milk allow the milk to stand for a few hours before making your cheese and hand skim some of the cream that has risen to the surface. You can then use this for cream cheese or for cooking. The exception to this is Stilton, for which high butterfat milk is used and, although this method has largely died out, it was once made from whole milk with extra cream added. Medium to low-fat cheese is also possible using high butterfat milk. To achieve this the milk should be thoroughly hand-skimmed before use or put through a separator and adjusted to give about 2 per cent butterfat.

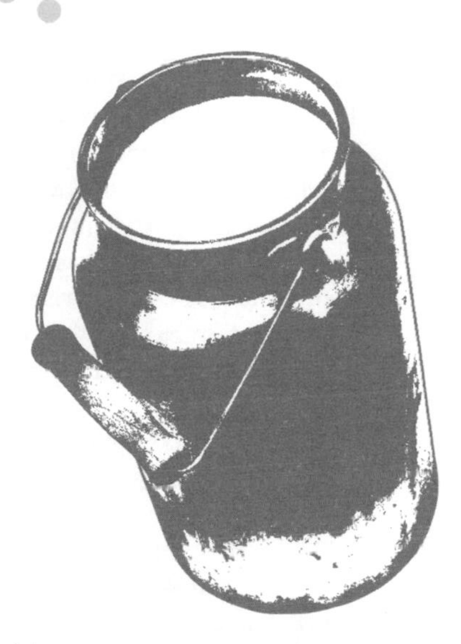

Raw milk

Raw milk from any outside source should be heat treated before use, particularly if the cheese is to be sold (always remember to check the regulations for the country in which the cheese will be sold). This will have already been done if you buy pasteurized milk; otherwise either raise the temperature of the milk to 68° C (154° F) and hold at that temperature for 30 minutes, or alternatively raise the temperature to 72° C (161° F) and hold for 30 seconds. This should, if at all possible, be done in a water-jacketed vessel to prevent the milk from 'catching' and should immediately be cooled, usually with cold water, to the temperature required by the recipe before adding starter.

Other milk types

GOAT milk for cheese making should be heat treated but never homogenized. All pouring, stirring and other movement of the milk should be done very gently. Milk that has been poured from a height or stirred vigorously releases the substances that cause the characteristic 'goaty' smell and flavour that some people dislike. The butterfat in goat milk has smaller fat globules which are held in the solids and less likely to rise to the surface. There is no need to hand skim goats' milk, even that with a high butterfat content, as you are unlikely to loose much fat to the whey.

SHEEP milk makes excellent cheese. It is high in butterfat, but like goat milk the fat globules are small and well supported by the solids. The solid content is also high and so the yield of curd per litre/pint will be higher than from cows' milk. However, the curd must be worked carefully to retain the water otherwise a dry cheese will result. Bag-curd and fast-drained cheeses are the best for sheep milk; Wensleydale, Stilton and Caerphilly are good recipes to select. Sheep milk cheese can also be blue veined with care and good storage, but the soft cheeses are less suitable. Other than simple acid curd they tend to be very firm and rather tough. Heat treatment of sheep milk should always be the high temperature, short time method. Slow heating and long holding tend to precipitate some of the protein and make renneting difficult.

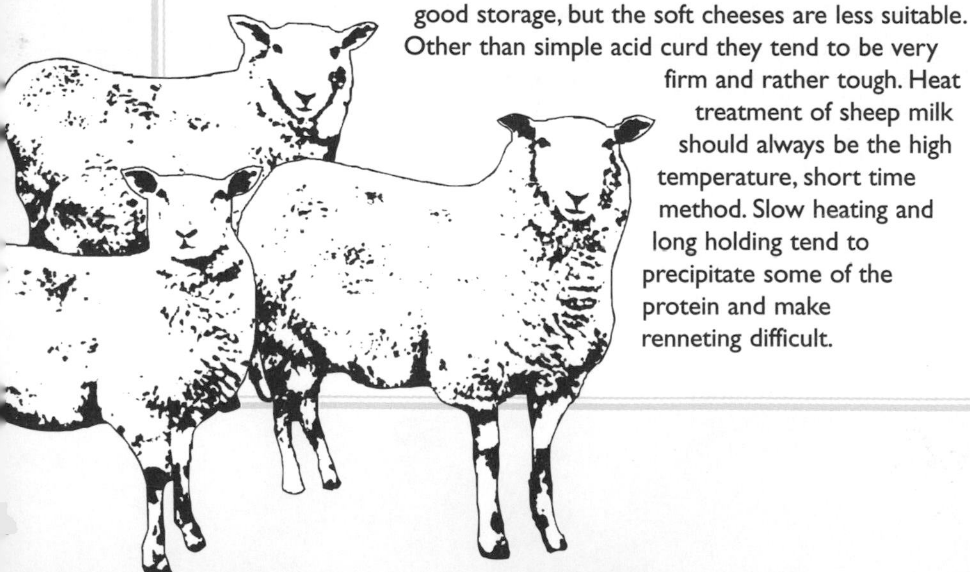

BUFFALO milk is handled similarly to cows' milk, and lends itself best to cheeses such as Mozzarella that are traditionally made using this type of milk.

SOYA milk is not true 'milk' but is used to make a form of cheese, mainly for use by vegans and the lactose intolerant. Apart from simple curd cheese, the production of soya cheese is best left to professionals as it requires certain additives and some scientific knowledge to get a palatable hard cheese.

UHT milk can be used to make cheese but certain additives are necessary to get a good coagulation necessary for hard cheese. If you are planning on making cheese using UHT milk it would be worth researching this further and getting some advice on the additives needed (see Suppliers on page 125).

DRIED MILK POWDER, suitably reconstituted, can be used but as most powdered milk is skimmed you are limited to soft cheeses and even then may need to add some cream.

FROZEN MILK, once thawed, can be used in exactly the same way as fresh milk. Milk should be frozen as soon as possible after milking. Cool the milk first and then pour into plastic bags placed inside 1 or 2 litre (1¾ or 3½ pints) plastic containers and freeze. When solid throughout, the bagged milk can be removed from the containers and stacked in the freezer and the containers reused. Do not use containers bigger than 1 or 2 litres (1¾ or 3½ pints) as large blocks will take too long to defrost and some precipitation of protein and fat will occur during thawing.

Equipment

Simple soft cheese can be made with the minimum of equipment, most of which you may already have at home, but molded soft cheeses, hard and semi-hard cheeses will require some specialist tools (see Suppliers on page 125).

To make simple soft cheese you will need:

Thermometer to monitor the temperature of the milk (preferably digital).
Wide bowl to set the curd.
Ladle with a small shallow bowl and a cutting edge, suited to the size of your mold.
Second container or bowl in which to drain the curd.
Cheesecloth squares to strain the curd through.
Clothes pegs to hold the cheesecloth together while the curd is draining.

To make molded soft cheeses, hard and semi-hard cheeses you will also need:

Vat or main vessel large enough to hold the amount of milk you are planning to use with a minimum of 5 cm (2 in) space above the milk (to help prevent spillage). Do bear in mind that it is difficult to work small amounts of milk in a vat that is too big. Ideally the vat should have a water jacket, i.e. one vessel inside another. A double saucepan, a bain marie, or even a bowl in a sink of hot water will do. It helps when cutting the curd to have a square or oblong vat, but one soon learns to cut in round or oval ones.

Molds made from food-quality plastic or stainless steel (it is not advisable to use coloured plastic pipe or tin plate). Coulommier molds are usually in two parts to make turning the cheese easier, but they are not essential. Soft cheese molds have no holes in the body of the mold as they drain from the bottom, but molds for hard cheese are perforated. Hard cheese molds also have loose plates top and bottom, making it easier to remove the cheese after pressing.

Weights to apply pressure to the top plate of the mold. This is essential when making hard cheese. Well washed stones can be used or small cement blocks can be made by putting a plastic bag into an empty mold and setting 5 or 6 cm (2 or 2½ in) of cement in it. Any of these options will enable you to increase or decrease pressure as required. Various designs of press can be bought (see Suppliers on page 125) using springs and/or levers.

Knives are essential for cutting the hard cheese curd. Special knives are made for this purpose (see Suppliers on page 125) but you can always improvise by using a pallet knife for your vertical cuts and a bent, stiff wire to cut horizontally.

Small pots or cups to contain the rennet and water, and also for dipping the whey.

Sieve for pouring the whey through in order to salvage the small particles and return them to the vat.

Bucket to hold whey which, incidentally, should never be poured down the drain. It can be put to use on your compost heap, poured onto the grass, fed back to stock or used in pig food.

Boards and **mats** to stand soft moulded cheeses on to drain. Boards should be the right size for the molds and not too big otherwise turning becomes impossible. For domestic use these boards can be wood or plastic. Matting is available from dairy suppliers but you could use cane table mats or pieces of fine colourless plastic mesh from a gardening centre.

Cloths, lots of them in various sizes. Natural cheesecloth can be bought in some fabric shops or you can use old, thin, cotton sheeting. Heavy materials can slow down drainage and mark the surface of your cheeses so avoid these.

Tablespoon for measuring ingredients.

Teaspoon for measuring ingredients.

Small hypodermic syringe is useful for measuring small amounts of rennet.

Kitchen knife, about the size of a dinner knife, to scrape down the soft curd on the cloth and to cut the curd blocks.

Ingredients

Cheese making requires relatively few ingredients. Other than milk (see pages 34–37), starter, rennet, salt, fat or oil and flavouring are usually all that is needed.

Starter

All modern cheeses need laboratory-produced starter cultures, the selected bacteria for cheese making, to ripen the milk. Even the small occasional producer will have to purchase this, but with careful management it need not be expensive. There are two main types of starter culture available from suppliers; simple cheese culture and DVI (direct to vat inoculation). DVI needs no pre-preparation and, as the name suggests, goes straight from the sachet to the vat of warm milk. The drawback with DVI for the small producer is that the smallest sachet available contains sufficient for 50 litres (88 pints) of milk and it is not advisable to portion the dry powder as this can affect the balance of types of bacteria in the culture. So the small producer will have to use simple cheese culture and pre-prepare it.

Preparing the starter

At least 8 hours before you wish to make your cheese bring 1 litre (1¾ pints) of milk to the boil, cool to 25° C (80° F) in a lidded container, add the starter powder and mix well in. The powder tends to float and a small whisk is useful to mix it into the milk. The inoculated milk is then left in a warm place until the milk is sufficiently acid to coagulate to the consistency of set yogurt. After using the amount required by the recipe any excess starter can be frozen, in measured quantities, for future use. If it is frozen in ice cube bags it will remain uncontaminated. One ice cube is sufficient to turn a litre (1¾ pints) of milk into a Coulommier and 8 cubes will make a litre (1¾ pints) of fresh starter for other cheeses. Do not repeat the process of freezing the surplus, as this risks contamination and weakening of the balance of organisms in the starter. Strain it through a cloth and use as acid curd cheese instead.

Rennet

As it is no longer feasible, not to mention particularly safe, to dry a calf or lamb stomach and soak strips of this in warm milk, rennet can be bought from dairy suppliers (see page 125). Rennet is sold either as a liquid or a powder. It is more economical to use liquid rennet for small-scale production, and, provided that it is kept refrigerated, a bottle of rennet will keep well. You can buy either animal or microbial rennet. Animal rennet comes from calf stomachs and is not suitable for goat or sheep cheese made for the bovine allergic or for vegetarians. However, microbial rennet is suitable for all types of milk and vegetarians too. It is also worth noting that sometimes a bottle of rennet will seem weaker than another. It is quite alright to increase the amount used in future makings, up to twice that recommended in the recipe, to achieve a firmer set.

Salt

Most cheese recipes include salt, added either during processing or by brining or dry salting the cheese after pressing or molding. Whilst special dairy salt is ideal, it is perfectly alright to use good cooking salt. But do not use table salt; there is a 'free running additive' in all domestic salt to stop it forming lumps and table salt contains more of this chemical and will give a bitter flavour to the cheese.

Fat or oil

Hard cheeses that are not brined or surface salted are rubbed with fat or oil after pressing. This keeps the surface supple and prevents cracking. Use butter, solid vegetable fat or olive oil. Coat the cheese all over and be prepared to re-grease it as cracks appear, especially if the cheese has needed a lot of cleaning.

Flavourings

If you are adding herbs, fruit, wine etc. to your cheese, ensure that these are mould free. Herbs and fruit should be thoroughly washed and drained before chopping and adding. It is better to use fresh herbs where possible as dried herbs can contain mould spores. Beer, wine or cider should also be pasteurized before use.

Working and storage areas

Before you begin to make cheese you will need to think about your working and storage areas. A well planned working area and a clean, well ventilated storage area will help your cheese making to be a success.

Working area

Any reasonable sized kitchen or utility room is suitable for small-scale domestic production. You will need a heat source (probably a stove) unless you have a self heating vat, and a source of hot and cold water. Try to work on a table or worktop at a height that allows you to reach the bottom of your vat easily. You will spill milk and whey however careful you are, so old sheets or large towels are useful both on the work surface and the surrounding floor area. It is, however, possible to improvise with your working area. Cheese has been made satisfactorily on a camping stove in a tent, and in some parts of the world it is made under the most primitive conditions. As long as you have heat, water and a clean environment you can go ahead.

Cheeses need somewhere to drain. Bagged curd cheese can be pegged in the cloth to the side of a bowl or bucket above the line of draining whey, whilst molded curd cheeses should stand on mats and boards. If stood in large baking tins with rigid bottoms, the whole set-up can be moved without too much disturbance. The whey can be removed by soaking it up with cloths. Hard cheeses in the mold can also be stood on trays to catch the drips.

Storage

Hard cheese requires careful storage for several weeks once it is removed from the mold. The storage area must be clean, well ventilated and neither too hot nor too cold. A temperature of about 12° C (54° F) is ideal. It should not be too dry, but this can be overcome by covering cheeses with damp cloths. The cheeses should stand on matting or slatted racks so that the bottoms of the cheeses can be kept dry. If you have several cheeses in store, a wire pan rack or vegetable rack is useful and can be covered with a damp pillowcase. The cheeses will grow mould on the rind and need daily cleaning with clean paper kitchen towels, as well as turning top to bottom. If the mould growth is sticky or shiny, do not be afraid to give the cheese a good wash under a cold tap, dry it well and re-grease or re-salt if necessary. Blue veined cheeses require an even lower temperature. A refrigerator set at 8° C (46° F) is good storage, although you need to keep the cheeses moist. White moulded cheeses are best kept in a refrigerator inside plastic bags. Open the bags daily and turn the cheese over. Be careful not to store blue veined and white moulded cheeses together.

How to make cheese

Before you attempt any of the recipes in this book, read the following basic method carefully to familiarize yourself with the process. While each of the recipes here vary depending on the type of cheese being produced, most cheese making involves the same principles.

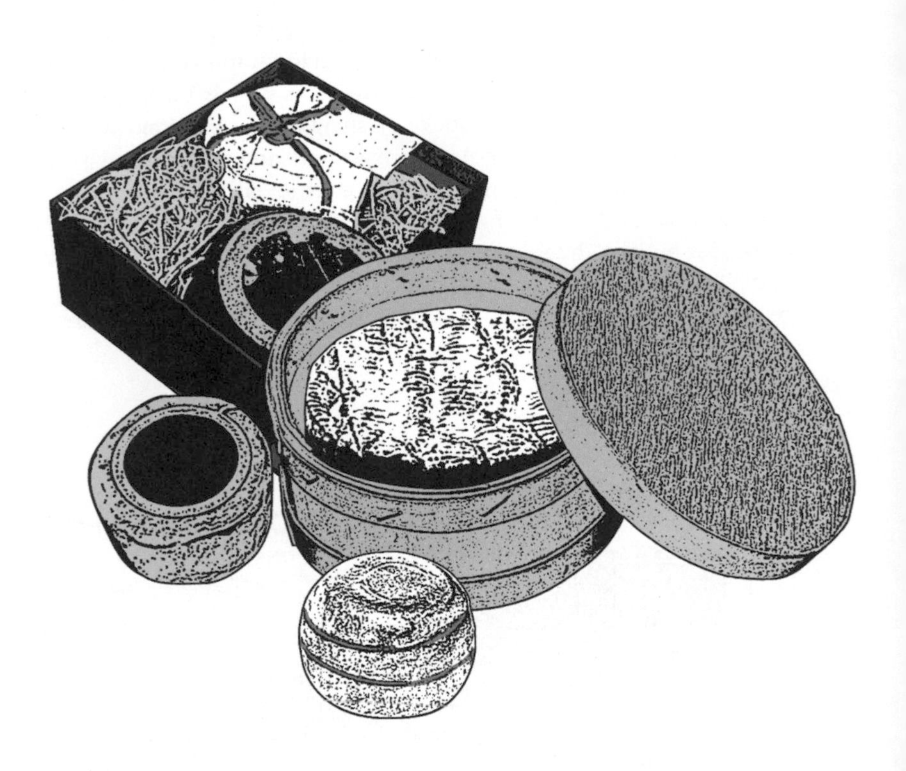

Prepare the milk and starter. If using frozen milk this must be removed from the freezer at least 24 hours before making your cheese. Remove the block(s) from the plastic bags and put them into clean containers with covers to keep out any contaminants (see page 37 for more on using frozen milk). Liquid milk should be kept refrigerated until use. The starter should be prepared at least eight hours before use, unless you are using frozen cubes straight from the freezer for smaller quantities of soft cheese (see pages 41–42).

Clean your work area and hands. Thoroughly clean your work area and put away all other foodstuffs, particularly those containing yeasts and those with strong smells such as garlic and onion that may taint the cheese. Make sure all of your equipment is to hand and thoroughly cleaned. Wash your hands with non-perfumed soap or detergent and clean your fingernails (fingernails are best kept short as the action of acid and enzymes softens nails and they can break off in the curd).

Heat the milk. Put warm water into the lower part or water jacket of your double vessel or vat up to the level at which it will not spill over when the milk vessel is inserted. Put the milk vessel into the water jacket and measure in your milk. Raise the temperature of the milk by heating the water jacket by whatever means available to you, to that required by the recipe. Always raise the temperature slowly – if you heat it too fast some of the protein (albumin) will precipitate, making small lumps in the curd that may stick to the vessel.

Add the starter. Use only the amount specified in the recipe as there is no advantage in over-inoculation of the milk. Rub the starter through a sieve to remove any lumps which will not disperse throughout the milk and may leave wet, over-acid pockets in the curd. Stir well. You can use a spoon or paddle for this purpose but your hand is by far the most sensitive tool for the job. Acidification or ripening will begin as soon as the starter is added to the warm milk. Keep checking the temperature and keep it at that required by the recipe until ready for the next step. The time the milk needs to ripen will be variable. When making large commercial quantities of cheese, the degree of ripening is constantly tested with an acid or pH meter to ensure that the final level is the same in every batch. This is not essential on a domestic scale. When the ripening time given in the recipe has elapsed and/or the stirred milk on the back of your hand smells of melted butter or slightly sour, it is time to add the rennet.

Add the rennet. Rennet should be measured into a small cup, jug or bowl in accordance with the recipe. It should then always be diluted with 4 to 6 times its volume of cold water. This is to enable the rennet to mix evenly throughout the ripened milk. Spread the diluted rennet all over the surface of the milk and immediately stir it in. Keep stirring until you are sure it is thoroughly mixed but not for too long as there is no way of guessing when coagulation will begin and you do not want to break the gel.

Top stir. Top stirring is carried out to stop the butterfat in the milk rising to the top and being lost to the whey after cutting or ladling the curd. This could lead to a tough, dry cheese. To top stir, dip the fingers of one hand into the milk to about the first joint, about 2.5 cm (1 in), say. Now make slow, gentle movements to keep the top of the milk moving. Do not make circles or move fast or you will end up making butter instead of sinking the cream. When your fingers leave a wake or trail in the milk, stop stirring immediately and leave to set. It is not important to top stir homogenized milk as the process itself breaks down the size of the fat globules so that the cream does not rise to the surface. When the setting time is up, test the curd for coagulation. Stick your little finger into the curd and curl it up out of the milk. Keep the hole you make as small as possible. If the curd breaks leaving milk on your finger and in the hole you have made, leave the curd a further ten minutes and try again in a different part of the vessel. The curd is ready to cut when you get a 'clean split' – your finger is free of milk on removal from the curd and the hole does not fill with milk.

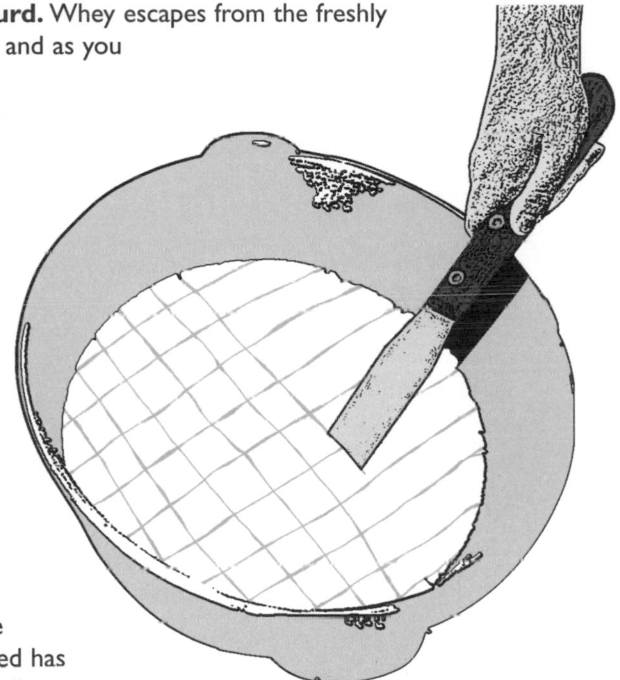

Cut or ladle the curd. Whey escapes from the freshly cut faces of the curd and as you need a good, even drainage, the curd should be cut or ladled to give as much surface area as possible. If the recipe calls for the latter method, ladle in long thin strips taken from side to side of the vessel and lay them carefully into the mold or cloth. Continue until all the curd that can be ladled has been taken out. If small particles are left, scrape them carefully together and put them in the middle of the filled mold. If cutting the curd in the vessel, cut in straight lines from left to right and then back to front. Use a pallet knife or a vertical cheese knife for this. Cut evenly and do not damage the curd any more than you have to when inserting or withdrawing the knife from the curd. Your vessel will now contain pillars of curd from which whey will start to run. Quickly cut these pillars into small cubes with a horizontal cheese knife or a wire bent at a right angle. The horizontal knife should be inserted down the side of the vessel and moved smoothly from side to side or front to back until all the curd is cut. The bent wire should be lowered a little into the curd and pulled across, lowered a little without withdrawing and pulled back, and so on until the bottom of the vessel is reached. If the vessel is not too wide, the wire can be shaped to the width and only one pass will be needed at each level. After cutting is complete, stir the curd to keep the pieces separate but do this gently so that the curd is not broken any smaller.

Scald the curd. Raise the temperature of the curds and whey according to the recipe. This raise in temperature causes the curd pieces to shrink and expel more whey. Time and temperature vary from cheese to cheese and have a noticeable effect on the texture and storage quality of the cheese. Scalding is complete when you can gently squeeze a small handful of curd without it breaking up.

Pitch the curd. This allows the curd to sink to the bottom of the whey in the vessel where it will form a mat. This mat of curd can be pushed to one side of the vessel, either flat or sloping depending on the instructions given in the recipe.

Drawing the whey. Unless your vessel or vat has a whey drain, remove the whey with a small jug or bowl. There will be small pieces of curd

floating in the whey, so pour it into a bucket through a sieve and return the retained curd to the curd mat. A flat-sided plastic box is useful for removing the whey from the bottom of the vessel and the last dregs can be soaked up with a clean cheesecloth. Keep the curd warm during this drawing off period or the whey will stop running and you will have a wet curd.

Cut into blocks. The curd should be cut with a knife into blocks of the size and shape specified in the recipe. Stack the blocks or put into cloths. If the blocks are to be stacked, build them around the sides or one end of the vessel so that the whey can run freely. Raise the vessel at the end where the curd is stacked so that it does not sit in the whey. Remove the whey as it collects. If the curd is to be bagged at this stage, follow the instructions given in the recipe. The blocks or bags of curd are pulled apart and re-stacked until the curd reaches the required texture as described in the particular recipe.

Mill the curd. This means to break or cut the curd into small pieces before packing. Curd can be torn or cut by hand, or if making larger quantities, fed through a curd mill. This process is not to aid drainage but rather to ensure a smooth texture without gaps in the final cheese. Small jagged pieces bond more readily than large blocks.

Pack the curd into molds. Line cheese molds with clean cloths before packing with curd unless otherwise instructed. Cheese not in cloths can be difficult to remove from the mold for turning. Pack carefully, pushing down into the mold with your knuckles rather than your fingertips which tend to crush the curd. Also make sure the cloth is not trapped in the curd, as this will leave a gap in the finished cheese where unwanted moulds could grow. Some whey will run from the curd whilst packing, so stand your mold in a tray or bowl.

Apply pressure. This is not always required, but when it is called for, apply just enough to make the whey run and no more. Too much pressure immediately after packing will squeeze out the butterfat and make a dry cheese. It will also stick the cheesecloth to the curd and you will tear the skin of the cheese when removing the cloth at turning. Pressure can be increased after the first turn and then daily until the removal of the cheese from the mold. Un-pressed cheeses need careful handling until they have formed a firm skin.

Finish your cheese. Salting, coating or brining can take place when the cheese has been removed from the mold for the last time. The cheese can be rubbed with dry salt, coated all over in softened fat, rubbed with oil or dropped into brine as the recipe requires. Be sure that, whichever method is used, the whole cheese is treated and salt, fat or oil is repeatedly applied as necessary on a day-to-day basis throughout the maturing period.

Store the cheese. The finished cheese should be stored in a cool and damp place. If the atmosphere is dry, cover the cheese with damp cloths. Turn and clean daily. Remove any mould with dry paper kitchen towels and discard after use. If the mould growth is excessive or sticky it does no harm to scrub the cheese lightly with a brush or pan-scrub whilst holding the cheese under the running cold tap. After any treatment, make sure that the cheese is thoroughly dried and be sure to fill any surface cracks with fat or salt.

Recipes

The recipes in this chapter range from Curd Cheese, the simplest form of soft cheese, to the more complex hard cheeses such as Cheddar and Dunlop. Once you have mastered the soft and hard cheeses here you may like to try some of the recipes and flavouring ideas in Taking Cheese Further on pages 86–101, remembering to refer to the Troubleshooting section on pages 82–83 should you run into difficulties.

Soft cheeses

What could be more enjoyable than tucking into some delicious home made Cream Cheese or serving up a cheesecake made with your very own Curd Cheese to family and friends. Soft cheeses are generally simpler to make that hard cheeses (although there are exceptions to the rule) so here is a good place to begin your cheese making.

Curd cheese

Curd cheese is the simplest form of cheese. It can be used for savoury snacks, in cooking and for making cheesecakes. It is ideal for smaller amounts of milk and can be made with high- or low-fat milk.

Method

1. Raise the temperature of the milk to 32° C (90° F). Add starter – 150 ml (10 tbsp) to 5 litres (8¾ pints) of milk. Stir well and leave until a firm curd has formed, maintaining the temperature throughout. This will take several hours.
2. Ladle the curd into a clean cloth lining a bowl or bucket. Leave for 15 minutes.
3. Draw the corners of the cloth together to form a bag and suspend over the bowl to drain. Do not let the bag hang in the whey. Open the bag occasionally during draining and scrape down the curd with a knife to unblock the weave of the cloth.
4. After a few hours, the curd should be drained sufficiently to remove from the bag as a lump. It should be soft and moist but with no visible whey running from it. Add salt to taste if the cheese is for savoury use.
5. Pack into containers or wrap in film and refrigerate. Use within 7–10 days.

Cheesecake
Curd cheese is ideal for making cheesecake. Follow the above recipe to make 1 litre (1¾ pints) of milk into curd cheese. Make a biscuit base with crushed digestive biscuits and butter or bake a thin sponge base in an 18 cm (7 in) diameter baking tin with a removable base. Put the curd cheese, 30 ml (2 tbsp) of caster sugar, a pinch of ginger powder, the juice of half a lemon or orange and a little finely grated rind into a food processor and mix to a smooth paste. Spread the mixture onto the base, decorate to taste and refrigerate for at least 1 hour before serving.

Cottage cheese.

Cottage cheese has a mild flavour making it ideal for cooking as well as enjoying with a green salad or cooked potatoes. It can be made using milk of different fat levels but if using a very low fat milk it is advisable to stir in a little cream at the end to enhance the texture.

Method

1. Raise the temperature of the milk to 21° C (70° F). Add starter – 6 ml (a little more than a tsp) per litre (1¾ pints) of milk. Immediately add rennet – 1 ml (¼ tsp) per litre (1¾ pints) of milk diluted in 4 times its volume of water. Stir well and leave until a firm curd is formed – this can take several hours.
2. Cut the curd into cubes, first cutting both ways with a vertical knife and then one way with the horizontal knife or wire. Heat the curd in the whey, stirring all the time, until the temperature reaches 45° C (113° F).
3. Quickly drain through a cloth and discard the hot whey. Tip the curd from the bag into very cold water and stir until the curd is cold. Carefully break up the clumps of curd by hand whilst in the water. Drain through the cloth again.
4. Return the curd to the cold vat and add 1 gram of salt per original measurement of milk if required and/or any other flavouring. If you have used a very low fat milk, stir in a small amount of thick cream. Spoon the cottage cheese into pots.

Serving suggestion

Stuff boneless chicken breasts with cottage cheese mixed with chopped and cooked mushrooms and garlic. Dip in egg, coat with breadcrumbs and fry gently on all sides until the chicken is cooked thoroughly.

Coulommier

> **Serving suggestion**
>
> Coulommier is lovely in a salad and can also be cooked – oil the cheese and fry lightly on both sides in a spicy butter until golden.

Coulommier is a traditional cheese drained in a round mold on a mat with a board beneath. It can be made in any shape of mold but should be kept small – not more than 15 cm (6 inches) in diameter – or it will be too unwieldy to turn. A purpose-made Coulommier mold is in two parts, the upper ring being about half the height of the lower. They can be made in stainless steel or plastic, the latter being the more stable. Both rings are filled initially, the upper part being removed when the curd has sunk into the lower ring. This double mold is not an essential item but it is more convenient (see Suppliers on page 125).

Method

1. Raise the temperature of the milk to 32° C (90° F). Add starter – 5 ml (1 tsp) per 1 litre (1¾ pints) of milk. Stir well. Add rennet – approximately 1 ml (¼ tsp) per litre (1¾ pints) of milk diluted in 4 times its volume of cold water. Stir well and leave until a curd has formed that gives a clean split over your finger.
2. Put your mat and board into a suitable dish or tray and ladle the curd in thin slices (see page 49–50), placing them carefully into the mold. Try not to break up the slices of curd as this will prevent the cheese from draining properly. When all the curd is in the mold, cover with a small piece of cheesecloth to take the whey off the top.
3. Remove the whey from the dish regularly as soon as it reaches the level of the board. Leave to drain until the whey has ceased or almost ceased to run.
4. Place a clean mat followed by a clean board on the top of the mold and, grasping the mold sandwiched between the two mats and boards in both hands, turn the whole thing upside down. Leave for 1 hour and then turn again (if the cheese sticks to the mat, use a blunt knife to ease it off and rearrange the top layer of curd). Leave for 24 hours.
5. The next day, when the whey has stopped running and the cheese is loose in the mold, lift it off the mat and lightly salt the cheese all over as required. The whole draining process is best carried out in a warm environment.

Colwick cheese

This is a fresh soft cheese molded into the shape of a bowl and can be made as small individual cheeses or a larger one for family use. The finished bowl can be salted if it is to be filled with savoury foods or sprinkled with sugar if it is for a dessert.

Method

1. Raise the temperature of the milk to 32° C (90° F). Add starter – 1 ml (¼ tsp) per litre (1¾ pints) of milk. Stir well. Add rennet – 1.5 ml (¼ tsp) per litre (1¾ pints) of milk diluted in 4–6 times its volume of cold water. Top-stir until coagulation begins. Leave until a curd has formed that gives a clean split over your finger.
2. Line a mold with a square of cheesecloth just large enough to fill the mold with the edges hanging over the side. Stand the mold on a mat and board in a dish or tray and ladle the curd in thin slices (see page 49–50) into the mold until it is full. Lay the corners and edges of the cloth on top of the ladled curd and leave to drain for 1 hour, removing the whey from the tray before it covers the board upon which the mold stands.
3. Take the corners of the cloth and tie in a bundle, then place it in the centre of the cheese where it will sink into the curd and form a bowl. Remove whey as it collects in the tray as before.
4. When the bag of cheese is firm enough to handle, turn it over and replace it in the mold with the knot underneath. Leave for several hours and then carefully remove the cloth from the bowl-shaped curd and place it the right way up on the mat and board. Leave to dry.
5. Salt all over if the cheese is for a savoury dish, or sprinkle lightly with castor sugar if it is to be used for a sweet dish.

Serving suggestion

Fill a salted Colwick cheese bowl with chopped smoked ham or flaked smoked salmon and potato salad. Fill a sugared bowl with fresh berries or a fruit compot and top with whipped cream or cold custard.

Cream cheese

Cream cheese can be made with single or double cream. Single cream cheese is rennetted but no rennet is needed for double cream, which should form a fairly firm curd as it ripens.

Single cream cheese

1. Use single cream with a butterfat content of 20–25 per cent. Raise the temperature of the cream to 24° C (75° F). Add starter – 5 ml (1 tsp) per litre (1¾ pints) of cream. Leave to ripen for 2 hours. Add rennet – 2 ml (½ tsp) per litre (1¾ pints) of cream diluted in 4 times its volume of cold water. Stir well. Leave until a firm curd has formed – this may take several hours.
2. Ladle the curd into a cloth over a clean vessel and gather the corners together to form a bag. Suspend this bag over the vessel so that the bag does not sit in the draining whey. When the whey stops running from the bag, open the cloth and, using a blunt knife, scrape the curd to unblock the weave of the cloth. Do this at intervals over the draining period until no more whey is running. Scrape the cheese from the cloth bag into a bowl and mix well until a smooth, lump free texture is achieved.
3. Add 1 gram of salt per litre (1¾ pints) of cream if required and/or any other flavouring. Pack into containers and refrigerate for up to 10 days.

Suggested additives

Chives, or other well cleaned herbs, a small, chopped and crushed garlic, finely chopped spring onions, small pieces of smoked ham or smoked salmon or some finely grated strong hard cheese such as extra mature cheddar make an interesting addition. Some people add chopped, cooked fruit such as apricots and pineapple to good effect.

Double cream cheese

1. Use double cream with a butterfat content of 50–59 per cent. Slowly raise the temperature of the cream to 24° C (75° F). Do not heat too quickly or overheat, otherwise the fat in the cream will start to melt. Add starter – 5 ml (1 tsp) per litre (1¾ pints) of cream. Stir well. Leave until a firm curd has formed – this may take several hours.
2. Ladle the curd into a clean cloth lining a bowl or bucket. Draw the corners of the cloth together to form a bag and suspend over the bowl or bucket to drain. Remove the whey as it collects and do not let the bag hang in the whey.
3. When drainage slows down take the bag of curd, fold over the corners and lay in the bottom of the vessel. Put a board on top of the bag of curd and apply light pressure by means of a weight on top of the board. This should not weigh more than 1 kg (2 lb 4 oz). Leave for 2–3 hours.
4. Remove the cheese from the cloth and transfer to a clean bowl. Mix well until the cheese has reached a smooth texture and add salt or other flavourings. Pack into containers and refrigerate for up to 10 days checking regularly for signs of mould growth.

Serving suggestion

This cheese can be molded into shapes using pots, bowls or pastry cutters. Line your mold with greaseproof paper and smooth in the curd. Refrigerate for 24 hours and then remove from the mold. Decorate the cheese with herbs, olives or fruit or serve with a salad.

Hard cheeses

To make hard cheeses you will need a suitable vat or main vessel, molds, weights and some additional equipment listed on pages 38–40. See page 125 for a list of suppliers.

House cheese

A simple hard cheese ideal for the beginner that lends itself to experimentation. This recipe is so-called because it can easily be adapted to fit in with the maker's daily routine.

Method

1. Raise the temperature of the milk in the vat to 32° C (90° F). Add starter – 7.5 ml (1¾ tsp) per litre (1¾ pints) of milk. Stir well and leave for 40 minutes. Add rennet – 0.5 ml per litre (1¾ pints) of milk diluted in 4–6 times its volume of cold water. Stir well. Top stir if necessary and leave for 40 minutes until a curd has formed that gives a clean split over your finger.
2. Cut the curd both ways with the vertical knife and one way with the horizontal knife or wire to form small cubes. Stir gently. If uncut pieces of curd appear, cut them to size.
3. Raise the temperature of the curds and whey slowly to 38° C (100° F), stirring all the time. The cubes should become smaller and fairly firm to handle. Allow them to settle to the bottom of the vat for 10 minutes, then decant or drain the whey from the vat and push the mat of curd to one end. Keep removing the whey as it drains from the curd.
4. Cut the mat of curd into blocks of about 6 cm (2½ in) and stack around the sides of the vat. Leave to drain for 10 minutes, removing the whey as it collects. Pull or cut the blocks apart, turn them over and re-stack.
5. Continue stacking, turning and removing whey at 10 minute intervals until a block of curd can be squeezed gently without more than a few drops of whey running out, and when pulled apart the curd tears rather than crumbles. Mill or break the curd into walnut sized pieces and place back into the vat, turning the curd to prevent it sticking together.

6. Add 1 gram of salt per litre (1¾ pints) of milk used and mix thoroughly. Pack carefully into a mold or molds lined with damp cheesecloth. Apply just enough pressure to make the whey run and leave for 2 to 3 hours.
7. Turn the cheese upside down into a clean damp cloth and press again with a little more pressure than before. Increase the pressure as the cheese shrinks. Turn daily for 4 days, but on the last day return to the press without a cloth to remove the surface marks.
8. Remove from the press, coat the cheese with fat or oil and put into store. Turn daily and keep clean. This cheese can be eaten at 4 weeks or left up to 10 weeks to increase flavour.

Serving suggestion

This is a good cheese for making sandwiches and it also toasts well. Try it in cheese sauces, and any other cooking which calls for a hard cheese, for an excellent flavour.

Caerphilly

Caerphilly cheese originated from the area around the town of Caerphilly in Wales. It is thought that while the cheese wasn't made in Caerphilly itself, it was sold at the town's market and therefore became known by the same name.

Method

1. Raise the temperature of the milk in the vat to 21° C (70° F). Add starter – 15 ml (1 tbsp) per litre (1¾ pints) of milk. Stir well. Raise the temperature of the milk further to 32° C (90° F) and leave to ripen. This will take around 1½ hours.
2. Add rennet – 1 ml per 5 litres (8¾ pints) of milk diluted in 4–6 times its volume of cold water. Stir thoroughly and leave to form a firm curd (this may take up to 1 hour).
3. Cut the curd both ways with the vertical knife and one way with the horizontal knife or wire to form cubes. Stir gently.
4. Raise the temperature of the curds and whey to 34° C (93° F) and continue to stir over a period of 20–30 minutes. Allow the curd to settle in the whey for 10 minutes. Remove the surface whey and push the curd to one side of the vat.
5. Form the curd into a cone shaped pile, removing the whey as it collects, and then cut the pile into wedges. Stack the wedges at the side of the vat and cover with a cloth to keep them warm. Leave for an hour or so until the curd looks and feels silky but is still moist.
6. Remove one wedge of curd at a time and cut on a board with a sharp knife into 2.5 cm (1 in) cubes. Work quickly as it is important that the curd does not get cold. Return the cubes to the warm vat. Add a small pinch of salt per litre (1¾ pints) of milk used and mix thoroughly. Do not over-salt.
7. Pack the curd into a mold lined with a warm, damp cloth. Add a little pressure – just enough to consolidate the curd – and leave for two hours.
8. Turn the cheese top to bottom into a clean, warm, damp cloth and apply a little more pressure. Leave for 24 hours.
9. Remove the cloth and return the cheese to the mold. Apply medium pressure, leave for 4 hours and then remove the cheese from the mold.

10. Place the cheese into a container with a 20 per cent brine solution (made by dissolving 200 g/7 oz of salt in boiling water and adding cold water to make each litre of brine). The cheese should float in the cold brine, so make certain it remains totally immersed. Leave for 24 hours, remove from the brine and allow to dry before putting into store. The cheese will be ready to eat in three weeks.

Serving suggestion

This mild, clean cheese is best enjoyed as a table cheese and is excellent in sandwiches and salads.

Cheddar

Cheddar is probably the best known cheese throughout the Western world. Originally named after the village in Somerset, England, where it was first produced, it has been known since the latter part of the 16th century. It is now manufactured in huge quantities in America, Australia and New Zealand and in smaller quantities in many other countries. The name 'Cheddar' is no longer area specific but is applied now to describe stages in the process of manufacture. 'Cheddaring' is used to describe the blocking, cutting, piling and turning of the curd in a particular fashion and all Cheddar worthy of the name is treated in this way. In the UK the quality of Cheddar marketed is carefully controlled. Only that cheese which is judged to have graded highly enough is sold as Cheddar cheese. Any inferior cheese is either re-processed or used as an ingredient in cheese products. Originally and until not so long ago all Cheddar cheeses were capped and bandaged with fine cloth greased into the skin followed by a stitched outer layer of course, heavier linen before they were stored away until mature. Mould would grow on the surface of the cheese and on the bandages, creating the hard greyish rind which can still be seen on a traditionally-made farmhouse Cheddar today.

Cheddar is one of the most labour-intensive cheeses to make. Good hand-made Cheddar requires knowledge, care and practice. It is not a cheese with which to start your cheese making but you will undoubtedly, sooner or later, want to try your hand. A traditional full size Cheddar is cylindrical in shape and can weigh up to 32 kilograms (70 lb 8 oz), a 'truckle' is the same shape but weighs 4–5 kilograms (8 lb 12 oz – 11 lb) and a chedlet, or baby Cheddar, 1.5 kilograms (3 lb 5 oz) or less. It is recommended that the home cheese maker should limit their early efforts at making Cheddar to about 2 kilograms (4 lb 8 oz) using, say, 20 litres (35¼ pints) of milk.

Method

1. Raise the temperature of the milk in the vat to 29° C (84 ° F) whilst hand stirring gently. Add starter – 7–10 ml (1½–2 tsp) per litre (1¾ pints) of milk. Stir well and leave for 30–45 minutes, hand stirring occasionally and maintaining a temperature of 29° C (84° F). Add rennet – 0.5 ml per litre

(1¾ pints) of milk diluted in 4–6 times its volume of cold water. Stir well. (If using homogenized milk, simply stir in the diluted rennet and leave to coagulate. If using non-homogenized milk, stir in the diluted rennet and gently top-stir the surface of the milk with the tips of your fingers to stop the butterfat rising to the surface until coagulation begins). Leave for 40–45 minutes, maintaining the temperature, until a curd has formed that gives a clean split over your finger with no milkiness in the whey.

2. Cut the curd both ways with the vertical knife and one way with the horizontal knife or wire to form small cubes. Try not to crush the curd into the corners or around the edges of the vat when you are cutting.
3. Raise the temperature of the curds and whey slowly to 40° C (104° F) over a period of 30–40 minutes, hand stirring continuously. If uncut pieces of curd appear, cut them to size. The cubes should become smaller and fairly firm to handle. Allow them to settle to the bottom of the vat for 10–15 minutes.
4. Push the curd, either by hand or with a suitable board, to one end of the vat and hold it there for 5–10 minutes until it stays in position when the hands or board is removed. Decant or drain the whey from the vat without disturbing the curd.
5. Cut the mat of curd into blocks of about 10 cm (4 in) and stack around the sides of the vat. Leave to drain for 15–20 minutes, removing the whey as it collects. Cut the blocks, turn them over and re-stack.
6. Continue cutting, turning and removing the whey at 15–20 minute intervals for 1½–2 hours until the blocks are firm and silky and most of the whey has been removed. Mill or break the curd into walnut sized pieces and place back in the vat, turning the pieces frequently to prevent them sticking together. Do this as quickly as you can so that the curd does not get too cold. Stir the milled curd thoroughly.
7. Add 7 grams of salt per 5 litres (8¾ pints) of milk used and stir thoroughly. Pack carefully into a mold or molds lined with damp cheesecloth, pressing down with your knuckles as you do so. Apply medium pressure and leave in a warm room for 4–8 hours, being very careful not to over-press. If the curd has been thoroughly cheddared, very little whey should run during the pressing.
8. Turn the cheese upside down into a clean dry cloth. Return to the mold

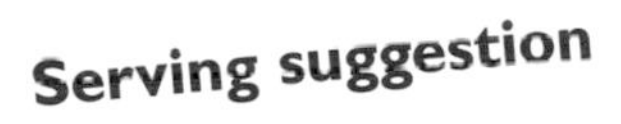

Serving suggestion

Cheddar cheese is a great all-round cheese and can be used for snacks and sandwiches or for sauces, soufflés, flans and dressings.

and apply maximum pressure. Turn daily for 4 days. On the fifth day return to the press without a cloth for about 4 hours to remove the surface marks.

9. Remove from the press, coat the cheese with butter or oil and put into store. Turn daily during the first week and at least weekly thereafter. Check for signs of cracking and fill any cracks with butter or oil. It is perfectly normal for your cheese to grow a dry, greyish-white mould which can be dusted off from time to time with a dry paper towel or cloth. However if wet or highly coloured moulds develop, these should be removed, once the skin has hardened, by scrubbing under cold running water with a bristle brush or a nylon pan-scrub. Dry thoroughly after washing and renew the grease coating. If there are traces of cheese mites, signalled by a fine dust round the cheeses in store, both the cheeses and the store must be thoroughly cleaned and all traces removed. If any infestation persists, professional advice should be sought.
10. Large Cheddars can be matured under good storage conditions for up to eighteen months. Small cheeses will lose too much weight and acquire a high proportion of rind over such a long period but a good mature cheese can be achieved with careful storage and care in up to six months. Even after 4 weeks the young Cheddar will have developed a good flavour.

Cheshire

Cheshire is thought to be the oldest known variety of English county cheeses. It was certainly recorded during Tudor times and history suggests that the Roman army of occupation commandeered much of the production of the time. Traditionally, the cheese was made to three different recipes to take into account the seasonal variations in the milk. Nowadays, factory production has been standardized and the differences in the milk in spring, summer, autumn and winter are minimal due to the controlled feeding of the dairy cows by augmenting the variable quality and quantity of their grazing with carefully compounded cake. Similarly, some traditional Cheshire cheese was coloured, originally with carrot juice and later with annatto, a tropical fruit derived dye. These days it is usually left an uncoloured pale cream.

Method

1. Raise the temperature of the milk in the vat to 29° C (84° F). Add starter – 15 ml (3 tsp) per litre (1¾ pints) of milk. Stir well and leave for 30–45 minutes, hand stirring occasionally and maintaining the temperature. Add rennet – 1 ml per litre (1¾ pints) of milk diluted in 4–6 times its volume of cold water. Stir well. (If using homogenized milk, simply stir in the diluted rennet and leave to coagulate. If using non-homogenized milk, stir in the diluted rennet and gently top-stir the surface of the milk with the tips of your fingers to stop the butterfat rising to the surface until coagulation begins). Leave for 40 minutes, maintaining the temperature, until a curd has formed that gives a clean split over your finger.
2. Cut the curd both ways with the vertical knife and one way with the horizontal knife or wire to form 1 cm (⅓ in) cubes. Do not overcut and try not to crush the curd into the corners or around the edges of the vat when you are cutting.
3. Raise the temperature of the curds and whey slowly to 34° C (93° F) over a period of 30 minutes, hand stirring continuously. Turn off the heat and allow the curd to settle to the bottom of the vat for 30 minutes.
4. Decant or drain the whey from the vat without disturbing the curd and soak up the last traces with a cloth. Cut the mat of curd into blocks of about 10 cm (4 in) and stack around the sides of the vat, leaving a channel

into which the whey can drain. Cover the blocks with a warm damp cloth and leave to drain for 20 minutes, removing the whey as it collects. Break the blocks apart, turn them over and re-stack.

5. Continue breaking the blocks, turning them and removing the whey at 20 minute intervals for 1–1½ hours until the blocks are firm and most of the whey has been removed. Add 1 gram of salt per litre (1¾ pints) of milk used and turn the blocks over to help distribution.
6. Mill or break the curd into walnut sized pieces and place back in the vat, turning the pieces frequently to prevent them sticking together. Try not to allow the pieces to crumble. Pack carefully into a mold or molds lined with clean dry cloths. Do not apply pressure and leave in a warm room for 4 hours. Turn the cheese upside down onto a clean dry cloth. Return to the mold and leave overnight, still without pressure.
7. The following day turn the cheese upside down onto a clean dry cloth, return to the mold and apply light pressure. After 4 hours increase the pressure to medium and after another 4 increase to maximum. Turn daily for 2 days, applying maximum pressure. On the third day return to the press without a cloth for about 2 hours to remove the surface marks.
8. Remove from the press, coat the cheese with butter or oil and put into store. Turn daily during the first week and at least weekly thereafter. Clean off any mould growth with a paper towel or cloth and re-grease or oil as necessary. The cheese is ready for use after 4 weeks but can be kept for 3–4 months in good storage conditions.

Serving suggestion

Cheshire is an ideal melting cheese for toppings and toasting. It is also a good dessert cheese to serve with apples or green grapes.

Derby

Derby cheese originated from the many dairy farms in the county of Derbyshire in central England. In the past it was never a standardised cheese in the sense that the method of production varied widely from area to area except in the physical shape and weight of the product. The cheeses were three times wider than tall, measuring about 40 cm (16 in) across and 13 cm (5 in) deep and weighed about 14 kg (31 lb). It was not until 1870 when factory production started that a uniform method and quality standards were introduced. Derby is usually an uncoloured cheese of uniform white to cream appearance. Its flavour should be clean and mild with a firm, moist and smooth texture with no holes or cracks on cutting.

Method

1. Raise the temperature of the milk in the vat to 21° C (70° F) whilst hand stirring gently. Add starter – 15 ml (3 tsp) per litre (1¾ pints) of milk. Stir well. Raise the temperature of the milk in the vat to 25° C (77° F) whilst hand stirring occasionally. Leave for 35–40 minutes. Add rennet – 2 ml per 5 litres (8¾ pints) of milk diluted in 4–6 times its volume of cold water. Stir well. (If using homogenized milk, simply stir in the diluted rennet and leave to coagulate. If using non-homogenized milk, stir in the diluted rennet and gently top-stir the surface of the milk with the tips of your fingers to stop the butterfat rising to the surface until coagulation begins). Leave for 35–40 minutes, maintaining the temperature, until a curd has formed that gives a clean split over your finger.
2. Cut the curd both ways with the vertical knife and one way with the horizontal knife or wire to form 1 cm (⅓ in) cubes. Leave for 5 minutes.
3. Raise the temperature of the curds and whey slowly to 35° C (95° F) over a period of 30 minutes, gently hand stirring continuously. Turn off the heat and allow the curd to settle to the bottom of the vat for 10–20 minutes until a mat of curd has formed on the bottom of the vat under the whey. Decant or drain the whey from the vat without disturbing the curd and soak up the last traces with a cloth.
4. Cut the mat of curd into blocks of about 10 cm (4 in) and stack around the sides of the vat, leaving a channel into which the whey can drain. Leave to drain for 20 minutes, removing the whey as it collects. Cut the blocks, turn them over and re-stack.

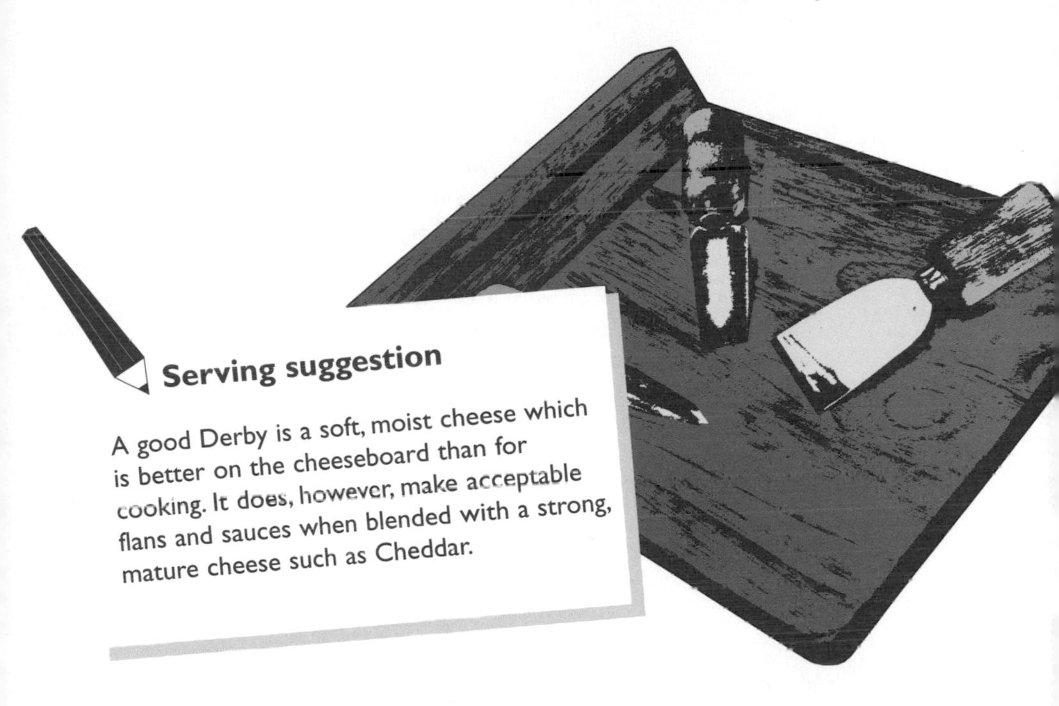

Serving suggestion

A good Derby is a soft, moist cheese which is better on the cheeseboard than for cooking. It does, however, make acceptable flans and sauces when blended with a strong, mature cheese such as Cheddar.

5. Continue cutting, turning and removing the whey at 20 minute intervals for 2 hours until the blocks are firm but still very moist. Mill or break the curd into cherry sized pieces and place back in the vat, turning the pieces frequently to prevent them sticking together.
6. Add 1.5 grams of salt per litre (1¾ pints) of milk used and stir thoroughly. Pack carefully into a mold or molds lined with damp cheesecloth, pressing down lightly with your knuckles as you do so. Fold over the cloth and add a rigid disc or follower. Apply just enough pressure to cause the whey to run, being very careful not to over-press, and leave in a warm room for 6–8 hours.
7. Turn the cheese upside down into a clean damp cloth. Return to the mold and apply medium pressure. The following day turn twice, once onto a clean dry cloth and once without a cloth, applying medium pressure on both occasions. Remove from the press and leave in a cool, well-ventilated room for 24 hours to surface dry.

8. Remove from the press, coat the cheese with butter or oil and put into store. Turn daily during the first week and twice weekly thereafter. Derby is a moist cheese and is susceptible to damage. Clean regularly during storage and check regularly for signs of cracking and mould growth and fill any cracks with butter or oil immediately. Dry, powdery grey or green moulds can be rubbed off with a dry cloth or paper towel. Sticky, wet or highly coloured moulds must be washed off with cold water, the cheese dried and then re-greased.
9. Derby is a mild cheese and ready for eating after as little as four weeks. It is not usually improved by keeping for more than twelve weeks.

Sage Derby

Sage Derby, a fashionable cheese for many years, is made in the same way as a Derby, but one third of the curd is separated after pitching and whey removal and is mixed with green colouring. This can be made by liquidizing and straining spinach, cabbage or other green vegetables, or by mincing and graining fresh sage. If sage is used, return some of the minced sage with the juice to the curd to give a pale green colour with small particles included. On filling into the mold, pack one third plain curd, pressed down with the knuckles, followed by one third sage and finally one third plain. This will give a clear green layer through the middle of the cheese.

Dunlop

There seems to be no other reason for the name Dunlop other than that it is from the village of that name, about 15 miles south west of Glasgow, in Ayrshire. The cheese is the Scottish version of Cheddar, originally made from the milk of Ayrshire cows. It was based on the traditional English Cheddar from Somerset, but, in common with other farmhouse cheeses, it was adapted to fit the working day of the farmers' wives and others who made it. It is therefore quicker to make and less time consuming than Cheddar.

Method

1. Raise the temperature of the milk in the vat to 30° C (86° F). Add starter – 10 ml (2 tsp) per litre (1¾ pints) of milk. Stir well and leave for 1 hour, maintaining a temperature of 30° C (86° F). Add rennet – 1.5 ml (just over ¼ tsp) per 5 litres (8¾ pints) of milk diluted in 5 times its volume of cold water. Stir well. (If using homogenized milk, simply stir in the diluted rennet and leave to coagulate. If using non-homogenized milk, stir in the diluted rennet and gently top-stir the surface of the milk with the tips of your fingers to stop the butterfat rising to the surface until coagulation begins). Leave for 40–60 minutes, maintaining the temperature, until a curd has formed that gives a clean split over your finger.
2. Cut the curd both ways with the vertical knife and one way with the horizontal knife or wire to form small cubes. Gently hand-stir the cubes in the whey for 10 minutes to prevent the cubes sticking together and encourage the whey to run from the curd.
3. Raise the temperature of the curds and whey slowly to 37° C (99° F), gently hand stirring continuously. Turn off the heat and allow the curd to cool for 10 minutes whilst hand stirring occasionally. Once the curds and whey have reached this temperature allow to cool whilst stirring gently for a further 10 minutes.
4. Allow the curd to settle to the bottom of the vat until a loose mat of curd has formed on the bottom of the vat under the whey. Push the mat of curd to one end of the vat with your hands and hold it there for 2 minutes or until it has formed a mass. Decant or drain the whey from the vat without disturbing the curd and soak up the last traces with a cloth.
5. Cut the mass of curd into pieces of about 10 cm (4 in) square and stack

Serving suggestion

Dunlop is a good multipurpose cheese. Use it as a table cheese, in sandwiches or for cooking, wherever a mild or medium hard cheese is called for.

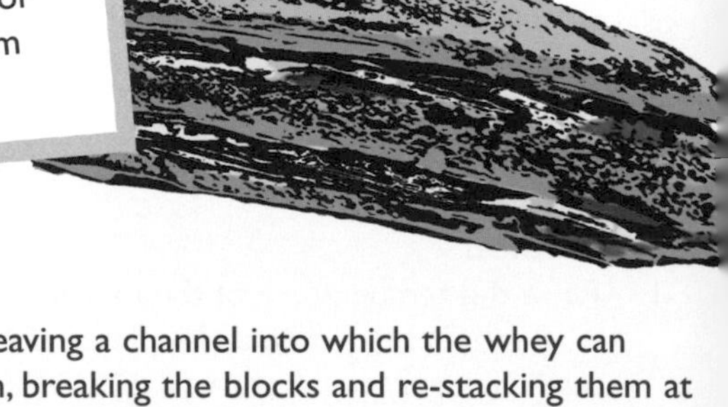

around the sides of the vat, leaving a channel into which the whey can drain. Allow the curd to drain, breaking the blocks and re-stacking them at regular intervals and removing the whey as it collects. Continue for about 30 minutes until the blocks are smooth and silky, and tear rather than breaking when pulled apart. Mill or break the curd into cherry sized pieces and place back in the vat. Stir well.

6. Add 5 grams of salt for every 5 litres (8¾ pints) of milk used and stir thoroughly. Leave for 5 minutes, retaining the warmth remaining in the vat, and then mix again. Pack carefully into a mold or molds lined with warm, damp cheesecloth, pressing down with your knuckles as you do so. Apply very light pressure for 15 minutes then increase to medium pressure and leave in a warm room for 4 hours.
7. Turn the cheese upside down onto a clean, warm and damp cloth. Return to the mold and apply maximum pressure for 24 hours.
8. Removed from the mold and cloth and immerse in water heated to 60° C (140° F) for 30 seconds, holding the cheese under the water to prevent it from floating to the surface. Remove the cheese from the water, dry the surface thoroughly and return to the mold lined with a fresh dry cloth and apply maximum pressure for 24 hours.
9. Remove from the mold and coat the cheese with butter or oil. Return it to the press without a cloth and apply maximum pressure for 4 hours to remove the surface marks. Put into store, turning daily during the first week and twice weekly thereafter. Check for mould regularly and clean the cheese as necessary. The sweet and mild cheese is best eaten after 8 weeks.

Gloucester

This cheese was originally prepared from the milk of the Gloucester breed of cow. Although they were renowned for the richness of their milk, they produced a poor quantity of it. Consequently, the breed was replaced by high yielders of less rich milk and the character of the Gloucester cheese altered as a result. Gloucester cheese came in two sizes, Single and Double, both of about 40 cm (16 in) in diameter, with Single being about 6 cm (2¼ in) deep and Double, twice that. Maturing times were different too, the smaller cheese being ready in six weeks but the Double Gloucester being at its best after six months to a year in store. Today, all the Gloucester commercially produced is Double which is lightly tinted (originally with carrot juice but now with Annatto) to give a warm, golden colour.

Method

1. Raise the temperature of the milk in the vat to 29° C (84° F). Add starter – 10 ml (2 tsp) per litre (1¾ pints) of milk. Stir well and leave for 1 hour, maintaining the temperature. Add the diluted Annatto a drop at a time until the desired colour is achieved. Add rennet – 1 ml (¼ tsp) per litre (1¾ pints) of milk diluted in 4–6 times its volume of cold water. Stir well and leave for 1 hour until a curd has formed that gives a clean split over your finger.
2. Cut the curd in both directions with the vertical knife and in both directions with the horizontal knife or wire. Stir gently to distribute

Serving suggestion

Gloucester has a very tasty, clean flavour which is excellent with apple or celery.

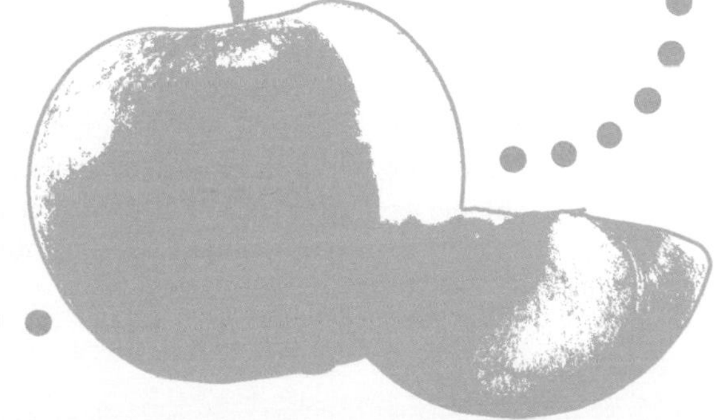

the curd. Cut the curd again in both directions with the vertical knife and in one direction with the horizontal knife or wire to form wheat-grain sized pieces. Gently stir the curds and whey for 10–15 minutes whilst maintaining a temperature of 29° C (84° F).

3. Raise the temperature of the curds and whey slowly to 35° C (95° F) over a period of 50–60 minutes, hand stirring continuously. Continue stirring for 10–15 minutes, maintaining the temperature. Turn off the heat and allow the curd to settle until a mat of curd has formed on the bottom of the vat under the whey. Push the mat of curd to one end of the vat with your hands and hold it there. Decant or drain the whey from the vat without disturbing the curd and soak up the last traces with a cloth.
4. Place a rigid plastic board or a shallow metal tray over the curd. Place a 1 kg (2 lb 4 oz) or so weight on top of the board or tray to assist consolidation and draining of the curd mass. Leave for 20–30 minutes, removing the whey as it collects by soaking it up with a cloth. Remove the board and weights.
5. Cut the mass of curd into pieces of about 10 cm (4 in) square and stack around the sides of the vat, leaving a channel into which they whey can collect. Allow the curd to drain, breaking the blocks and re-stacking them at regular intervals and removing the whey as it collects. Continue for about 1 hour until the blocks are firm and rubbery. Mill or break the curd into cherry sized pieces and place back in the vat.
6. Add 1.5 grams of salt per litre (1¾ pints) of milk used and stir thoroughly. Pack carefully into a mold or molds lined with damp cheesecloth. Apply medium pressure and leave for 24 hours. Turn the cheese upside down into a clean dry cloth. Return to the mold. Apply maximum pressure and leave for a further 24 hours.
7. Remove from the press and coat the cheese with butter or oil. Return to the press without a cloth for about 4 hours and apply medium pressure to remove any surface marks.
8. Ensure the cheese is well greased and put into store. Turn daily during the first week and at least weekly thereafter. Clean the cheese surfaces of mould as it appears and re-grease as necessary. Gloucester can be eaten as a mild cheese at 6 weeks old or matured for up to 12 months.

Wensleydale

Wensleydale was originally made from the milk of the ewes grazing on the Yorkshire moors. It can, however, be made from the milk of goats or cows or even mixed milk. The yield of cheese will be greatest from sheep milk as the solid content is very much higher than the alternatives. Wensleydale is a cheese which needs careful handling over a long period during making. You will need to plan ahead for a process spread intermittently over two days.

Serving suggestion

Wensleydale is widely useful in both sweet and savoury cooking. It is particularly tasty served with apple either as a desert or in a cheese and apple pie.

Method

1. Raise the temperature of the milk in the vat to 21° C (70° F). Add starter – 1–2 ml per litre (1¾ pints) of milk. Raise the temperature to 29° C (84° F) gradually over the next 20 minutes. Add rennet – 1 ml (¼ tsp) per 4 litres (7 pints) of milk diluted in 4–6 its volume of cold water. Stir well. If butterfat rises to the surface top stir until coagulation begins. Leave for 40 minutes, or until a firm curd has formed that gives a clean split over your finger, whilst maintaining the temperature.
2. Cut the curd both ways with a vertical knife to form pillars measuring around 1 cm (⅓ in) square. Move the curd very gently in the whey for 1 or 2 minutes using your hand, without stirring. This will prevent the pillars sticking together and so hindering drainage. Leave for 10 minutes, then cut one way with a horizontal knife or wire to form 1 cm (⅓ in) cubes. Stir thoroughly but gently until the cubes have separated and leave to settle for 20 minutes, maintaining the temperature in the vat.
3. Ladle the curd into a cloth suspended over a whey bucket to allow the curd to drain until it is firm. This can take as little as 2 hours or as long as overnight, and must be done in warm and moist conditions. The bag of curd must be suspended in such a way that it is clear of the draining whey and the bag and bucket placed in a warm environment.

4. Once the curd is firm turn it out of the cloth and return it to the warmed vat. Cut the curd into 10 cm (4 in) blocks. Place the blocks on a draining rack or board without stacking and mat at one end of the vat to allow any whey to escape. Leave for 1½ hours, turning the blocks top to bottom at 20 minute intervals.
5. When the curd is flaky, pulls apart without crumbling and is just moist when squeezed, break into walnut sized pieces. Add 1 gram of salt per 4 litres (7 pints) of milk and stir thoroughly. (Blue mould culture can be added at this stage, if liked, to achieve a blue veined Wendleydale. Add a small pinch of Penicillium Roquefort mould to the salt and mix well before adding the salt mixture to the curd.)
6. Pack the curd carefully into damp unlined molds and leave for 3 to 4 hours hours without applying pressure. By this time the curd should have gone together and it should be possible to turn it out in one piece. Turn out into warm moist cloths, return to the mold and leave for 24 hours in a warm environment, again without applying any pressure.
7. Remove from the mold, turn into a clean, damp and warm cloth and return to the mold applying medium pressure. Do not apply more than medium pressure. (This should be 10–12 kg/22–26 lb in weight when applied to a 15 cm/6 in diameter cheese, or the equivalent applied by spring or lever. Smaller diameter cheeses require proportionately less pressure, e.g. for 10 cm/4 in diameter cheese use a 5–6 kg/11–13 lb weight.) Leave for 8 hours.

8. Remove from the mold and cloth and cover well with fat or oil. Placing a cloth bandage around the cheese will help it to retain its cylindrical shape, but remove after the first week and re-grease.
9. Put into store, turning on its head and cleaning daily and re-greasing when necessary. This cheese can be used in 2 to 3 weeks but will develop a creamier flavour in up to 6 weeks. If the cheese is to be a blue Wensleydale, the cheese should be needled after the first week in store. It should be penetrated right to the centre of the cheese with, preferably, a stainless steel needle which will allow air into the cheese and so encourage vigorous growth of the blue mould. Needling should be repeated weekly until the cheese is ready for use. The blue variety will require at least six weeks cool storage before the mould will have spread throughout the cheese.

Troubleshooting

Soft cheese

Unsatisfactory texture

The key to good texture in soft cheeses is even and rapid draining of the curd. This requires that cheeses are put to drain in a warm place, such as is provided in warm kitchen or adjacent to a night storage heater. Some soft cheese makers use their vat or double saucepan with the water jacket kept warm for this purpose. Whatever you choose, do not let your draining curd get cold, as it will waterlog and take forever to drain. Good practice requires that, in the case of bagged curd, the cloth be scraped down, and for molded cheeses, they be turned regularly.

Cheese sticking to the mat

Coulommiers, especially, will stick to the mat in the early stages of draining. The mat is best removed by carefully rolling it away from the cheese on turning and repairing the damage with a knife or, better still, your finger.

Contamination

It is rare to get contamination of soft cheeses as long as the milk is fresh and all the equipment, your hands and the working area are spotlessly clean. Should your cheese become contaminated you should check your milk source and be careful to clean your hands and equipment thoroughly prior to making your cheese.

Hard cheese

The finished cheese is too wet
If, after storage, hard cheese is too wet, it points to the curd having been insufficiently worked and drained before packing.

Mis-shapen cheese
This indicates bad packing of the milled curd and/or inadequate pressure or time during pressing.

Holes in the cheese
Gassy or slimy holes usually indicate microbial contamination of the curd either in the milk, which should have been remedied by heat treatment, or during making. The slimy holes are most often caused by the presence of yeasts – to avoid this do not work in an area where bread or pizza are present.

The cheese is too dry
If the finished cheese turns out to be too dry, butterfat has probably been lost to the whey. Next time, be more careful with your top stirring, your control of temperatures and your milling and pressing.

Cheese mites
Hard cheeses, whilst in store, may be attacked by cheese mites. These are microscopic spider-like creatures that invade the rind of the cheese leaving behind a fine, sandy coloured dust on the shelves or boards upon which the cheeses are stored. The presence of mites will be detected before too much damage is done provided the cheeses are inspected daily. At the first signs of trouble, wash the cheeses in strong salty water, dry them on paper kitchen towels and return them to thoroughly cleaned shelves. Do not use insect sprays which may taint your cheese. If your cheese store becomes badly infested it will need a proprietary treatment obtainable from a pest control firm.

Taking cheese further

Once you have mastered soft and hard cheeses you can begin to think about taking your cheese making a stage further. Here you will find suggestions for personalizing your home made cheese by adding flavourings as well as more challenging recipes ranging from delicious soft moulded cheeses such as Brie to the strong blue veined Stilton.

Flavouring

Once you have mastered plain cheeses, you will want to progress to the more challenging varieties and a great way to start is by flavouring your cheeses.

Herbs and spices

Try making a Coulommier and, at the ladling stage, sprinkle fresh, washed and chopped herbs between each layer of curd. You can also roll the cheese in herbs once it is fully drained. Alternatively, rub the drained cheese all over in a paste of mashed garlic and salt. Wrap in a foil wrap and leave for several hours so that the garlic flavour permeates the whole cheese. You can also try adding spices of your choice. Coulommier or curd cheese with cinnamon and sugar go very well with fresh strawberries. You can, after molding and draining, drop your Coulommier or balls of curd cheese into wine that has been mixed with spices, heated and cooled. Leave for 12 hours, drain, dry and eat.

Liquid additives

Hard cheese can be flavoured and marbled with wine, beer, cider or apple juice. When the curd is milled and ready for salting, pour the heat-treated and cooled liquid over the curd pieces and turn them over in it. Let any excess liquid drain out of the curd before salting, packing and pressing. This will give an interesting flavour and a marbled appearance. All sorts of liquid additives can be used with cheese. It is a matter of taste – if it appeals to you, try it, but remember to heat treat any additive which might add yeasts to the curd.

Moulded cheeses

Once you have mastered flavouring your cheeses, the next advance in cheese making is to try making a soft, white moulded cheese like a Camembert or Brie. The white mould coat is created by coating the cheese with the white mould powder obtained from suppliers of cheese making sundries (see page 125). It comes in sachets that contain too large an inoculation for the small-scale producer, but they can be split up by mixing the contents of the sachet with 500 grams (1lb 2oz) of dry salt and then rubbing the coat of each cheese with about 2.5 grams ($^1/_{16}$ oz) of the mixture. The mould can also be suspended in water for spraying the coats, but it is probably best for the occasional producer to use the dry salt method as the mould will remain active in the unused salt for many months.

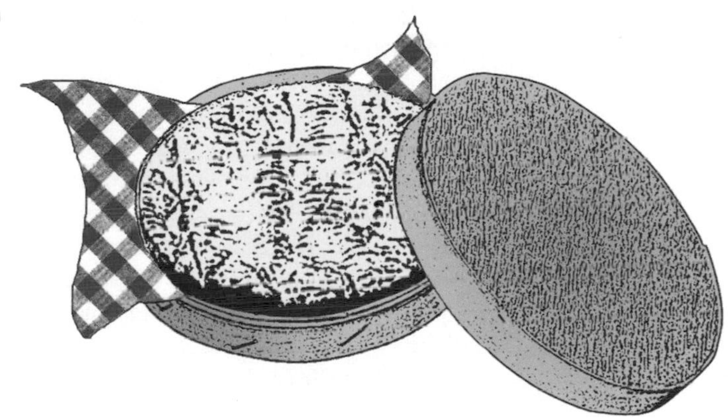

Note

Keep both Brie and Camembert away from blue veined cheeses to prevent cross contamination of the moulds that will spoil both types of cheese.

Camembert

This classic white moulded cheese is said to have originated in France during the French Revolution when its creator, an exiled priest, passed on the recipe to a resident of the village of Camembert.

Method

1. Raise the temperature of the milk to 29 ° C (84° F). Add starter – 0.5 ml (⅛ tsp) per litre (1¾ pints) of milk. Stir well and immediately add rennet – 0.5 ml per litre (1¾ pints) of milk diluted in 4–6 times its volume of cold water. Stir well and leave until a firm curd has formed. This could take some time as the rate of inoculation of starter is low.
2. Ladle the curd carefully in slices into plain molds without drainage holes, and stand each on a mat and board placed in a tray to collect the drainage. Allow the curd to settle, removing the whey as it collects, until the cheese can be turned in the mold with the aid of another mat and board placed on top. Continue turning at intervals until the cheese is firm enough to hold it's shape when the mold is removed.
3. Leave the cheese to dry for several hours and then rub the cheese all over with the salt and mould mixture or sprinkle with dry salt and spray with the suspension of mould in water. Put to store in a large plastic bag to give the cheese airspace at a temperature between 8 and 12 ° C (46 and 54° F). Turn daily.
4. Camembert is ready to use as soon as the mould has formed, but the flavour improves with ripening after about 6 to 8 weeks.

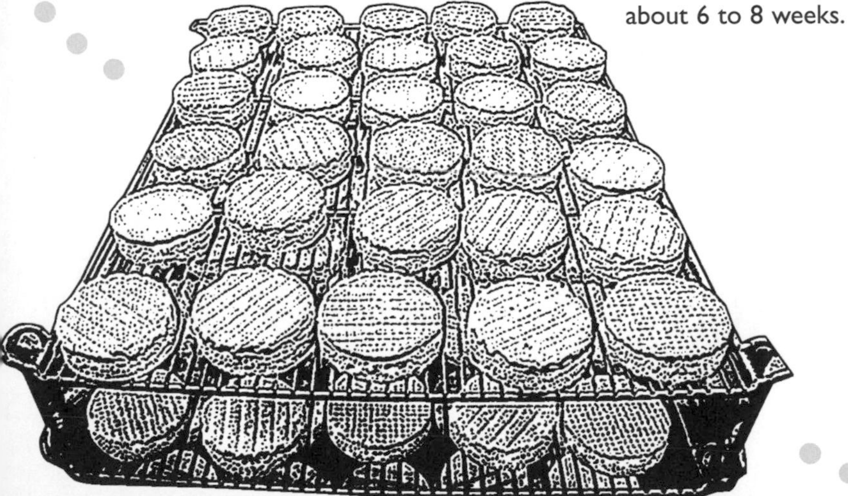

Brie

This cheese is very similar to Camembert in production, the main difference being in the shape of the mold – large diameter in relation to depth – which makes it difficult to turn. It will need to be very firm before the first turn is attempted and great care must be taken.

Method

1. Raise the temperature of the milk to 29 ° C (84° F). Add starter – 0.5 ml (⅛ tsp) per litre (1¾ pints) of milk. Stir well and immediately add rennet – 0.5 ml per litre (1¾ pints) of milk diluted in 4–6 times its volume of cold water. Stir well and leave until a firm curd has formed. This could take some time as the rate of inoculation of starter is low.
2. Ladle the curd carefully in slices into plain molds without drainage holes, and stand each on a mat and board placed in a tray to collect the drainage. Allow the curd to settle, removing the whey as it collects, until the cheese can be turned in the mold with the aid of another mat and board placed on top. Continue turning at intervals until the cheese is firm enough to hold its shape when the mold is removed.
3. Leave the cheese to dry for several hours and then rub the cheese all over with the salt and mould mixture or sprinkle with dry salt and spray with the suspension of mould in water. Put to store under a frame covered in damp cloths to encourage a moist atmosphere and ensure the temperature is around 15° C (59° F). Turn daily.
4. Brie is ready to use as soon as the mould has formed, but the flavour improves with ripening after about 6 to 8 weeks.

Blue veined cheeses

Blue veined cheeses are made by adding Penicillin Roquefort mould. Always refer to the instructions on the mould packaging but, as a rough guide, a sachet of Penicillin Roquefort from your supplier can usually be mixed with a kilogram (2 lb 4 oz) of salt and stored until needed. Most blue-veined cheeses are semi-hard as it is not easy to get blue veins to grow in the texture of soft molded cheeses as they do not admit enough air to encourage blue mould growth.

Stilton

To make Stilton you will need to be able to tie a Stilton knot. This is formed by drawing three corners of a piece of cheesecloth together and tying the remaining corner around them to make a slip knot which can then slide down to tighten the bundle.

Method

1. Raise the temperature of the milk to 29 ° C (84° F). Add starter – 0.5 ml (⅛ tsp) per litre (1¾ pints) of milk. Stir well and immediately add rennet – 0.5 ml (⅛ tsp) per litre (1¾ pints) of milk diluted in 4–6 times its volume of water. Stir well and leave for 90 minutes or until a firm curd has formed.
2. Cut the curd both ways with a vertical knife but do not cut horizontally. Allow to pitch (see page 50), letting the curd settle to the bottom of the vat in the whey for ten minutes. Now drain off or bale out the surface whey and ladle the curd carefully in slices into a cloth-lined bucket.
3. Fold the cloth corners over the curd in the bucket and leave in the whey for an hour. Tip out the whey and stand for a further hour to allow the whey to collect again. Tie the corners of the cloth of curd into a Stilton knot and remove from the bucket and stand the bundle on a draining rack. Tighten the bundle at hourly intervals until the curd is firm and only a little whey is running.
4. Turn the curd carefully out of the cloth into the warm vat and cut into 10 cm (4 in) blocks. Cover with a cloth to keep warm. Turn the blocks occasionally until the texture is flaky when broken.

5. Break or mill into walnut sized pieces and sprinkle with the salt and blue mould mixture, using 1.5 grams (1/16 oz) of blued salt per litre (1¾ pints) of milk. Mix thoroughly until dissolved.
6. Pack into molds without cloths adding a follower or fittings disc (see page 119) onto the top of the cheese. Do not apply any weights or pressure. Cover the mold with a damp cloth. Turn the mold daily for 4 to 6 days until the cheese can easily be removed.
7. Scrape the surface of the cheese with a blunt knife dipped in hot water until all of the surface cracks are filled. Put into storage and after a week pierce the cheese right through from side to side and top to bottom in about six places using a plastic coated steel needle or a knitting needle. This will allow air to enter the cheese and encourage the blue veins to grow.
8. The Stilton is ready for use after 2 to 4 months depending on its size.

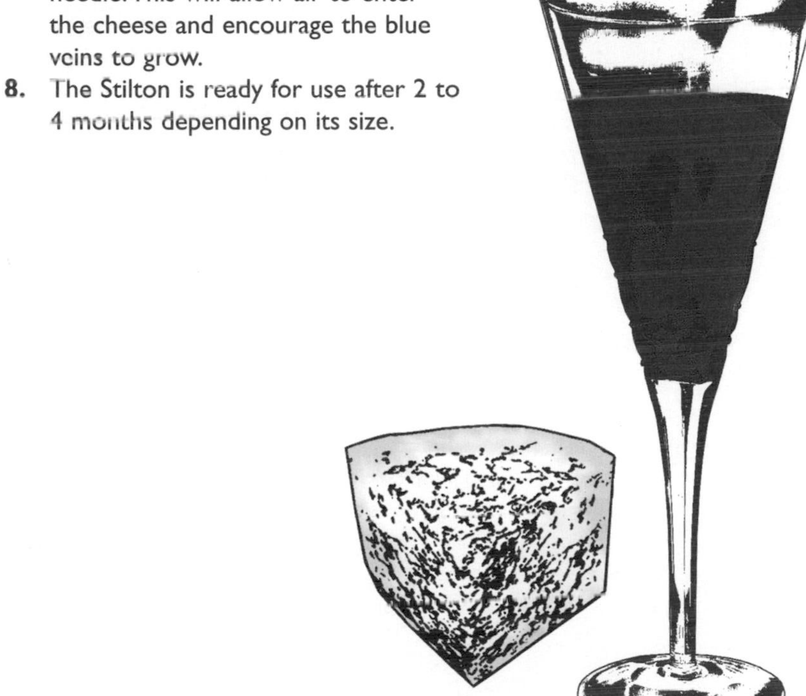

Dorset Blue

This cheese is traditionally made with milk that has been hand-skimmed to remove some of the fat but whole milk makes equally good cheese.

Method

1. Raise the temperature of the milk in the vat to 26 ° C (79° F). Add starter – 1 ml (¼ tsp) per litre (1¾ pints) of milk. Stir well and leave for 40–45 minutes to ripen. Add rennet – 0.5 ml per litre (1¾ pints) of milk diluted in 4–6 its volume of cold water. Stir well and leave for 1–2 hours until a firm curd has formed.
2. Cut the curd both ways with the vertical knife and one way with the horizontal knife or wire to form cubes. Stir gently for 5 minutes and then allow the curd to settle for 15 minutes.
3. Stir gently again for 5 minutes and then allow the curd to settle in the whey for 1 hour, ensuring that the temperature is maintained at 26° C (79° F).
4. Push the mat of curd to one side of the vat and remove the whey. Cut the curd into cubes of about 5 cm (2 in), bundle into a cloth and place on a draining rack. At 15 minute intervals over a period of 2 hours open the cloth, separate and turn the blocks, replacing them in the cloth each time.
5. The curd should now have a moist but not wet appearance and a flaky texture. Break or mill the curd into pieces the size of a small walnut and sprinkle with 1.5 grams of the blued salt per litre (1¾ pints) of milk.
6. Pack loosely into molds lined with a damp cloth and add just enough weight or pressure to make the whey run. Do not over-press. Leave overnight.
7. The following day remove the cheese from the mold and cloth and dip it into water at 60° C (140° F) for 30 seconds. Return the cheese to the press with the cloth and apply medium pressure for 24 hours. Remove the cheese from the press and cloth and rub thoroughly with dry salt.
8. Put into storage and after 2–3 weeks pierce the cheese right through from side to side and top to bottom in about six places using a stainless steel needle or a knitting needle. This will allow air to enter the cheese and encourage the blue veins to grow.
9. The Dorset Blue is ready for use after 12 weeks.

Cheeses of the world

While the principles of cheese making and the basic ingredients are the same throughout the world, the resulting cheeses are greatly varied (see pages 24–27). This variation is largely due to the type of animals kept in each particular region, the quantity of spare milk available and the amount of time the cheese maker had to devote to their cheese.

Feta

Feta is probably the best known Greek cheese and is a soft, brined cheese made from the milk of cows, goats or sheep or a mixture of any of these.

Method

1. Raise the temperature of the milk in the vat to 32° C (90° F). Add starter – 1 ml (¼ tsp) for every litre (1¾ pints) of milk. Stir well and leave for 40 minutes. Add rennet – 0.5 ml per litre (1¾ pints) of milk diluted in 4–6 times its volume of cold water. Stir well and leave for 50 minutes to 1 hour until a firm curd has formed.
2. Cut the curd into 3 cm (1¼ inch) cubes using a pallet knife for the vertical cuts both ways and a wire for the single horizontal cut. Leave for 15 minutes to pitch.
3. Drain through a cloth over a bucket and return the curd to the vat. Shape the curd into a large block using your hands and cover it with a clean cloth. Leave for 2 hours.
4. Cut the block into smaller 10 cm (4 inch) blocks and immerse them in a 20 per cent brine solution (see page 100) for 24 hours. Alternatively bury the blocks in salt for 2 days, turning once and re-salting if necessary.
5. Cut the cheese into strips and pack into plastic containers or seal in plastic bags and store for up to 30 days.

Halloumi

This Cypriot cheese is traditionally made from the milk of goats or sheep or a mixture of both. It is, however, also possible to make it at home using high fat cows' milk.

Method

1. Raise the temperature of the milk in the vat to 34° C (93° F). Traditionally no starter is used, but one or two drops per litre (1¾ pints) of milk will be useful if you are using pasteurized or heat treated milk. Stir well and immediately add rennet – 0.3 ml (⅛ tsp) per litre (1¾ pints) of milk diluted in 4–6 times its volume of cold water. Stir well and leave for about 45 minutes until a firm curd has formed.
2. Cut the curd into 3 cm (1¼ inch) cubes using a pallet knife for the vertical cuts both ways and a wire for the single horizontal cut. Stir gently whilst raising the temperature to 42° C (108° F). Continue to stir for a further 20 minutes.
3. Allow the curd to settle and then press the curd into a firm mat under the whey with the flat of your hand. Drain or ladle the whey from the vat and set aside. Cut the mat of curd into 10 cm (4 in) square blocks and pile them up in the vat. Cover with a cloth.
4. Put on a board on top of the pile, add some weight to aid drainage and leave for 2 hours. Meanwhile bring the whey to the boil in another vessel. Cut the blocks of curd into pieces measuring about 10 cm x 15 cm x 5 cm (4 in x 6 in x 2 in) and drop these pieces into the boiling whey.
5. Simmer for 45 minutes or until the blocks float in the hot whey. Remove the blocks from the hot whey, cool for about 20 minutes, sprinkle with salt and leave to cool completely.
6. Sprinkle herbs between the blocks, if liked, and pack into plastic containers or seal in plastic bags. The cheese can be eaten fresh or refrigerated for up to 2 weeks.

Mozzarella

To be authentic, this Italian cheese should be made from buffalo milk, but you can make it perfectly well from rich cows' milk. It is ideal for home made pizzas and other Italian recipes.

Method

1. Raise the temperature of the milk in the vat to 32° C (90° F). Add starter – 0.5 ml ($\frac{1}{8}$ tsp) per litre ($1\frac{3}{4}$ pints) of milk. Stir well and leave for 40 minutes to ripen. Add rennet – 0.3 ml for each litre ($1\frac{3}{4}$ pints) of milk diluted in 4–6 times its volume of water. Leave for 45 minutes until a firm curd has formed.
2. Cut the curd both ways with a vertical knife and one way with a horizontal knife or wire to form small cubes. Stir for 5 minutes and then leave to settle for 30 minutes without letting the vat or curd go cold.
3. Press the curd down and to one side of the vat with the flat of your hands. Drain or ladle the whey from the vat.
4. Cut the curd into 15 cm (6 in) squares. Wash the blocks in cold clean water and leave to drain.
5. Wrap the blocks in clean cloths and refrigerate for 2–4 hours. Immerse a block of curd in water heated to 82° C (180° F) until it is warmed through and can be stretched without breaking.
6. When the curd has tested satisfactorily, take the curd from the cloths and cut into ice-cube sized pieces. Immerse the pieces in hot water heated to between 72 and 82° C (162 and 180° F). Leave the water and curd to cool to 57° C (135° F) whilst kneading the curd in the water until a single, smooth, pliable, stretchy mass is formed.
7. Pack the hot curd into molds or bags or form into large balls. Immerse the curd (and the mold or bag, if using) in cold water for 1 hour to cool.
8. Immerse the curd (and the mold or bag, if using) in a 20 per cent brine solution (see page 100) for 12 hours.
9. The cheese can then be eaten fresh or vacuum packed and stored in a refrigerator for up to 6 weeks.

Panir or Surati

This un-pressed, white curd cheese is made throughout India and originated in what was once the Bombay province. Whey from the previous day's production was used as the starter and lemon juice as a substitute for rennet. Modern Panir or Surati is now made from rich cows' milk or mixed milk, ideally pasteurized or heat treated before use, but it was traditionally made from buffalo milk. Baskets, usually made of bamboo, are used to drain the curd. These vary in size from tea to dinner plate but are all quite shallow. They are prepared by rubbing their dampened inner surfaces all over with salt.

Method

1. Raise the temperature of the milk in the vat to 32° C (90° F). Add starter – 5 ml (1 tsp) per litre (1¾ pints) of milk. Stir well and immediately add rennet – 1 ml for each litre (1¾ pints) of milk diluted in 4–6 times its volume of water. Leave for 2 hours until a firm curd has formed, maintaining the temperature.
2. Ladle the curd into a pre-prepared basket in thin, even slices, laying one slice carefully on top of another until the basket is full. Place the baskets over a draining tray so that the whey can be retained and leave to drain in a warm place (traditionally the curd is placed in the sun to drain).
3. Once the curd has formed one solid mass and can be removed from the basket in one piece, pour the retained whey into a suitable container and carefully add the curd. Leave for 12–36 hours, maintaining a temperature of 25° C (77° F). The cheese may then be used straight away or cut into cubes and added to curries and soups.

Ricotta

Ricotta is an Italian cheese which is now produced in many countries. These include Africa, where, in Tanzania, it is part of a total milk usage programme where all of the milk is utilised including the water! The milk is first processed into a semi hard cheese. The whey from this is treated so that Ricotta can be recovered and is then boiled and reduced in volume to create Mysost, another whey cheese. During the prolonged boiling, the steam is condensed on corrugated roofing sheets and the distilled water is collected for use in baby food. Ricotta is prepared from the protein, mainly albumin, left in the whey after other cheeses have been made. The whey from Cheddar cheese production, for instance, contains fat, lactose, some vitamins and minerals and residual proteins, including albumin, which rennet does not coagulate. The first whey draining should be used for the production of Ricotta. Later drainings, particularly those after milling and salting contain salt and curd particles which do not dissolve and give a bitty, salty Ricotta.

Method

1. Filter the whey through a sieve or cloth and pour into a saucepan. (A jacketed vat is unsuitable for this purpose unless it is steam heated).
2. Measure a volume of milk equal to 5–10 per cent of the total volume of the whey and add this to the filtered whey. This will help to improve the final texture of the Ricotta.
3. Gently heat the whey and milk mixture until a few bubbles begin to appear (but not to boiling point) and the heat-coagulated protein begins to rise to the surface. Skim off the protein with a perforated spoon and put into a mold.
4. Apply light pressure until the mixture is cool and firm enough to handle. Rub a little salt into the surface, if liked, and leave the cheese to dry. Ricotta can be used fresh or salted as a base for sauces or unsalted as a topping for deserts. It can also be dried further and cut into cubes to be added to soups and casseroles.

Edam

The town of Edam is about 13 miles (21 km) north of Amsterdam and it gave its name to a cheese which is probably the most well known of the Dutch cheeses and the most widely produced. The spread of the cheese throughout the world has led to other names for the same product in other countries where Tete de Maure (France), Manbollen (Netherlands), and Katzenkopf (Germany) are typical. Edam is a cows' milk cheese, originally made from full cream milk, but today the fat is reduced to a content of about 2.5 per cent. This means that the home producer can use lightly skimmed whole milk, using the cream for other purposes, or whole milk if it has been homogenized. Milk sold as semi-skimmed is normally of 1.7 per cent butterfat and is not rich enough to give a good Edam.

Instantly recognized by its red waxed coat and spherical shape, the cheese requires special molds for its manufacture. Made in plastic, these are now readily available from suppliers in several sizes.

Method

1. Raise the temperature of the milk in the vat to 30° C (86° F). Dilute the annatto in a little water and add to the milk a drop at a time until the milk is a warm yellow colour (you can also use pasteurized carrot juice if liked).
2. Add starter – 15 ml (1½ tsp) for every litre (1¾ pints) of milk used. Stir well. Immediately add rennet – 0.5 ml (⅛ tsp) per litre (1¾ pints) of milk diluted in 4–6 times its volume of cold water. Stir thoroughly but gently for 3 minutes, then leave for 30 minutes, maintaining a temperature of 30° C (86° F) until a firm curd has formed that gives a clean split over your finger (it may be necessary to top-stir from time to time to prevent any butterfat from rising to the surface).
3. Cut the curd one way only with the vertical knife. Leave until the whey begins to run and accumulate between the slices and around the edges of the vat. Now cut the other way with the vertical knife and one way with the horizontal knife or wire. Any curd that has clung to the sides of the vessel and remained uncut should be carefully dislodged by hand, and then the vertical knife is passed through the curd in both directions until it has been cut into tiny uniform pieces about the size of cereal grains.
4. Raise the temperature of the curds and whey to 35° C (95° F), hand stirring continuously. Turn off the heat and continue stirring until the grains of curd are firm, then allow the curd to settle to the bottom of the vat. Decant or drain the whey from the vat without disturbing the curd and retain the whey.
5. Place the mold or molds onto a drainer. Squeeze the curd lightly in your hands to remove any excess whey before pressing firmly into the mold using your knuckles. Apply light pressure and either place the mold in the vat to keep warm or leave in a warm room for 30 minutes.
6. Raise the temperature of the retained whey to 60° C (140° F). Remove the cheeses from their molds and submerge in the hot whey for 2 minutes. Allow the cheeses to drain before returning them to the mold or molds lined with clean, warm, damp cloths. Apply medium pressure for 6–12 hours depending on the size of the cheese (larger cheeses will need to be pressed for longer).

7. Remove the cheeses from the molds. Either rub salt directly into the skin of the cheese or roll the cheese in salt. Return the cheese to the mold and apply light pressure. Turn and salt daily for 6 days, each time applying light pressure. Alternatively the cheese can be placed in a 20 per cent brine solution made by dissolving 200 g (7 oz) salt in a little hot water and adding cold water to make up to 1 litre (1¾ pint). Leave the cheeses in the brine for 6–7 days, turning daily.
8. Remove the cheeses from the mold or brine and rinse thoroughly in warm water, removing any slime or surface blemishes. Allow the cheese to dry before putting into store on shelves in a well-ventilated room and maintain a temperature of 15° C (59° F). Because of the shape of Edam cheeses, they must be stored on shelves in such a manner that any flattening is confined to top and bottom. Consequently, some form of support is necessary, a collar of rolled cheesecloth for instance.
9. Turn the cheeses daily for the first month and twice weekly thereafter. During each turning smooth the cheese all over with your hands to help maintain its shape. If slime mould appears on the surface, the cheeses should be washed carefully in tepid water and dried thoroughly. At about eight weeks old, the cheeses are cleaned, dried and red-waxed, or they can be rubbed with good oil to prevent cracking and wrapped in film or vacuum packed.

A note on wax

The red waxed coat is applied to Edam as a form of product identity and packaging for export. It is not obligatory in Holland and the home cheese maker can make a perfectly satisfactory Edam without the shape or the coat. The waxing is a difficult, expensive and time consuming exercise and is seldom wholly successful on a domestic scale. The special cheese wax is available from sundries suppliers in several colours and the coating is done by rolling or dipping in molten wax which has to be kept hot enough to avoid the cheese cooling the wax to setting point, but hot enough to melt the cheese.

Selling cheese

Selling your home-made cheese can be wonderfully rewarding but you will need to be properly equipped. This chapter provides advice on starting a small cheese business but it is always advisable to check the regulations and guidelines for the country in which the cheese will be sold to ensure you can meet with the regulations before you begin production.

Setting up

If you have enjoyed making cheese, if your family and friends enjoy the cheese that you make, or if a glut of home-produced milk is readily available, you may consider making some cheese to sell. This will involve a lot of pre-planning and a not-inconsiderable financial outlay on premises and for equipment and supplies. There are strict regulations and many guidelines pertaining to milk producers and processors, and you would be well advised as a first step to contact your local environmental health and trading standards offices to ascertain what is expected of you.

It is time and money well spent to take advice on any cheese making enterprise, but be sure to seek an advisor who has experience in this field. There are cheese making courses available at colleges and private organisations and there may also be a visitor facility at your local cheese factory, which could have a viewing gallery. The scale of production may be much bigger than for what you have in mind, but the principle is the same.

Before you begin to make cheese on a commercial scale, ask yourself the following questions:

What cheese or cheeses are you going to make?
What quantity of milk are you planning to use?
How will the milk be sourced?
Where are you going to make the cheese?
What equipment, tools and supplies will you need and where will you buy them?
Where are you going to store the cheeses after making?
What are you going to call your product?
How will you package and distribute your goods?
Where and to whom will you sell your cheese?

What cheese or cheeses are you going to make?

It is probably best to concentrate on making a cheese that you like yourself. It is difficult to make and judge a cheese as good if you don't like the taste of it! It's a good idea to make your favourite specimen cheese and offer it to others for comment. If general opinion concurs with your own then take that cheese as your starting point, but remember to choose a product that can be produced to fit around your daily timetable (see page 29). If you decide to make more than one variety always remember not to let blue and white moulded cheeses come into contact with each other or with non-moulded varieties – the moulds will contaminate all cheese made or stored in the same area or store.

What quantity of milk are you planning to use?

The volume of milk you process will depend on the amount readily available and the pairs of hands you can muster to process it, but remember, once you have to supply a market you will need to make a regular supply of cheese. One person in one day can easily process 100 litres (176 pints) of milk into cheese, maybe 125 litres (220 pints) if the vat and equipment are efficient. But the difficulty may arise when it is time for milling the curd and filling into molds or presses. Any delay will mat and cool the curd so speed is essential.

It is not a good idea to make large batches of soft-ladled curd cheeses single-handedly as the curd ladled last from the vat will be surrounded by too much whey for careful ladling. Any more than batches of 5 to 7 litres (8¾ to 12½ pints) of milk is a two person task when making Coulommier, Colwick or similar soft cheeses. Of course your market may dictate what you make but do bear in mind the practical constraints and plan accordingly.

How will the milk be sourced?

If you keep lactating animals then the answer to this question is simple. However you will need to ensure that the milk is handled and heat treated in such a way as to meet with your country's regulations. If you need to buy your milk from an outside source local dairy farms may be a good place to start, as are dairy wholesalers.

Where are you going to make the cheese?

A major step is to decide where the cheese is to be made. It is generally not permitted to produce food for sale in a domestic kitchen, but check your local food safety regulations for guidance. It is likely that you will need a cheese making room and a cheese store, the actual size of which will depend upon your projected production and, to a certain extent, on the type of cheese or cheeses you plan for.

A typical rectangular, purpose built vat to hold 125 litres (220 pints) measures approximately 100 cm long by 50 cm wide (40 in by 20 in) but it could be circular or oval in shape. Also remember to allow room for a pasteurizer if you are intending to heat-treat raw milk. The cheese making room should have adequate supplies of hot and cold water on tap, good drainage, impervious floor and walls, good lighting and controlled ventilation. If your vat is electrically heated and requires more than 3 kilowatts of electricity a standard plug socket is inadequate and you will need to arrange for an alternative electricity supply. And if you are using a heat source other than electricity it must be finely controllable.

You will also need a table or work-surface, which must be impervious to water on all sides and edges, an area for storing equipment and an area for draining and pressing prior to storage. If you are making soft cheese do bear in mind that its production takes up a lot of space and you will need a large, easily accessible draining table area where the cheese can be turned without spillage.

The provision of a large sink or dairy wash-up big enough to take molds, buckets and large items as well as a hand washing facility separate from the main sink are imperatives. The whole area must be easy to clean and disinfect and there should be space to hang up essential aprons, head coverings and protective clothing that must be worn during production. There should also be a place for cheese-room footwear to be kept, which should never be worn elsewhere.

You will find that the provision of a domestic washing up machine in which to wash molds and anything else that will fit in is a great help. The modern tablets or powders will dissolve protein during the wash cycle and nothing is as effective in cleaning and sterilizing curd knives. Your local regulations may not allow you to launder your protective clothing within the food production area but your domestic washer situated elsewhere should be adequate.

What equipment, tools and supplies will you need and where will you buy them?

Having planned your working area you will need to acquire the equipment and tools necessary for making cheese or cheeses of your choice. In addition to the following you will still need the assortment of measuring jugs, bowls, ladles and small tools that you will have used for domestic cheese making, together with a good supply of cheesecloths, either of washable cotton or disposable material. For those cheese makers who are going to make soft molded cheeses or other ladled curd, in quantities more than one person can handle, remember you will need enough ladles for each person and enough boards, mats and cloths for the production.

Vat

The ideal vat is oblong or square in shape and either has legs or can be positioned on a work top to bring it to a comfortable working height. It should be a water-jacketed vat with provision for heating the water, either internally with an immersion heater or with pumped hot water from an external source. The advantage of the second option is that the water jacket needs to be only a little larger than the liner in which the cheese is made and so contains less water to be heated and cooled as required.

The external hot water source can be a small tank or pipe containing heaters through which the water is pumped with an ordinary, domestic central heating circulating pump. For a 100 litre (176 pint) capacity vat, one 3 kilowatt heater is adequate provided you do not plan to use the vat to batch pasteurize your milk. The heater can be controlled by a simple energy controller as used on a domestic electric cooker. These are not true thermostats that regulate the actual temperature but they allow timed and repeated bursts of energy into the heater according to the dial setting.

The vat is best fitted with a whey drain and a jacket water drain, the latter fitted with a simple hose connection so that the jacket can be filled with water as required. The whey drain must never be too small; a

2.5 cm (1 in) clear valve is essential and a form of strainer, either before or after the valve, will save the loss of small particles of curd. The best vats are constructed with a slight fall from the sides to centre and from one end to the whey drain, to effect complete drainage.

There is no real alternative to a stainless steel vat if you are making cheese for the market and you need efficiency and longevity. However 'stainless' steel is a misnomer as although it will not rust it can stain and needs careful cleaning. Never use wire wool or even stainless steel scrubbers to remove hardened curd, only nylon that has not been impregnated with abrasive grit. Some larger producers use converted bulk milk tanks for their vats but these are frequently round or oval and are, without a cradle to raise them to working height, difficult to work in. Their shape makes a slight difference when cutting the curd, but one soon adjusts.

Knives

You will need, at the very least, a horizontal curd knife. A bent wire big enough to cut a large volume of curd would be unwieldy and give an uneven cut. Ideally a pair of American-style curd knives should be obtained with blade spacing and length to suit your product and your vat. There is an alternative available in the form of a single knife with diagonally set blades. The technique for cutting with a diagonally spaced knife is quite different and more difficult, but it can be mastered.

Buckets or holding tank
Several large buckets for holding whey or a pipeline to a holding tank will be required. A suitable arrangement will have to be made for disposal of the whey in an ecologically acceptable manner. While you can dispose of a domestic quantity of whey on your lawn, you certainly cannot flood the ground, drains or septic tank with many gallons.

Thermometer
Until the advent of digital thermometers, the floating mercury or alcohol in glass thermometer was found in every cheese-room. They had their obvious disadvantages because of the danger of breakage in the vat but they were accurate and convenient and didn't need batteries. Today you will need a digital probe thermometer, but it is advisable to compare your thermometer to others from time-to-time to make sure there are no discrepancies.

Molds and presses
There must be enough molds and/or presses available to accommodate the cheese being made on the day and any that is left over from a previous make. If you are making a soft, molded cheese which doesn't require any pressure to be added, there are a wide variety of light-weight plastic molds in all shapes and sizes readily available from suppliers.

There is less choice when it comes to more robust molds suitable for using under pressure. It is not advisable to use lengths of drainpipe or any other pipe made from PVC. If you wish to make your own molds, choose a plastic that is safe to come into contact with food; ideally polypropylene, high-density polythene or low-density polythene. The latter has the lowest resistance to heat and is not suitable for every process. All these materials, in a range of sizes, can be purchased from specialist stockholders and can be cut, machined and drilled with ordinary wood or metalworking tools.

Where pressure is called for, the 'heap of clean stones' suggested for small-scale production is unsuitable for regular, larger quantities of manufacture. Pressure must be applied in a controlled and constant fashion so that as the cheese loses whey, and therefore volume, the pressure is maintained. There are several ways of doing this, the simplest being to use springs. This method was used in Victorian times when a self-contained, two-piece double spring press was marketed. This press was copied in recent years in modern materials but is no longer available except, perhaps, second-hand.

The Wheeler press is a double spring press that has been available for many years and is still very popular. It is complete with a mold but it may not be found suitable once production reaches a certain level. As production increases, it is worth considering a single large, double lever arm press or similar and stacking it with molds and followers to utilise its large capacity. These lever arm presses, or the similar spring loaded presses, occasionally come up at farm auctions and, if they are not snapped up by antique dealers, they can be refurbished and put to good use. If you are fortunate enough to find one of these presses, make sure that the cast iron dish upon which the molds sit is not cracked. This has usually been caused by outside storage when water has frozen in the dish.

Commercially there is another method of pressing (particularly hard pressing) cheese. This is done in a horizontal gang press where a single ram, either pneumatic or hydraulic, acts on a single stacked column of tapered molds, the base of one mold pressing into the top of the next just like a stack of paper cups. There are a dozen ways of applying pressure via levers, weights and springs and it is not difficult to construct a suitable rig yourself. Bear in mind that the materials used must be suitable for the purpose, however your choice may well be made for you by what you can purchase second hand. Cheese making equipment, by and large, wears well and if made properly in the first place can be bought second hand with care and confidence.

Curd mill

The tedious and time-consuming task of breaking up the curd by hand can be alleviated with a curd mill. There are power-driven mills available with very large throughputs and a much smaller hand operated peg mill with a throughput in excess of 5 kg (11 lb) per minute. There are the names of suppliers and manufacturers on page 125 to help you locate equipment, but do scan suitable farming and smallholding publications for second hand equipment.

Starter, rennet, salt and fat

Make certain that you always have adequate supplies of starter and rennet. Once you are using more than 40 litres (70 pints) of milk per batch, to make, for example, a Cheddar type cheese, you can use Direct Vat Inoculation (DVI) which avoids the pre-preparation of starter. Experiments may make this possible with other types of cheese, but some commercial cheese-makers still prefer to make up starter the day before with simple starter products.

Rennet can be bought in large quantities and safely refrigerated for some weeks. Salt can be bought in bulk and your local baker could advise on a source of white fat to coat your cheeses.

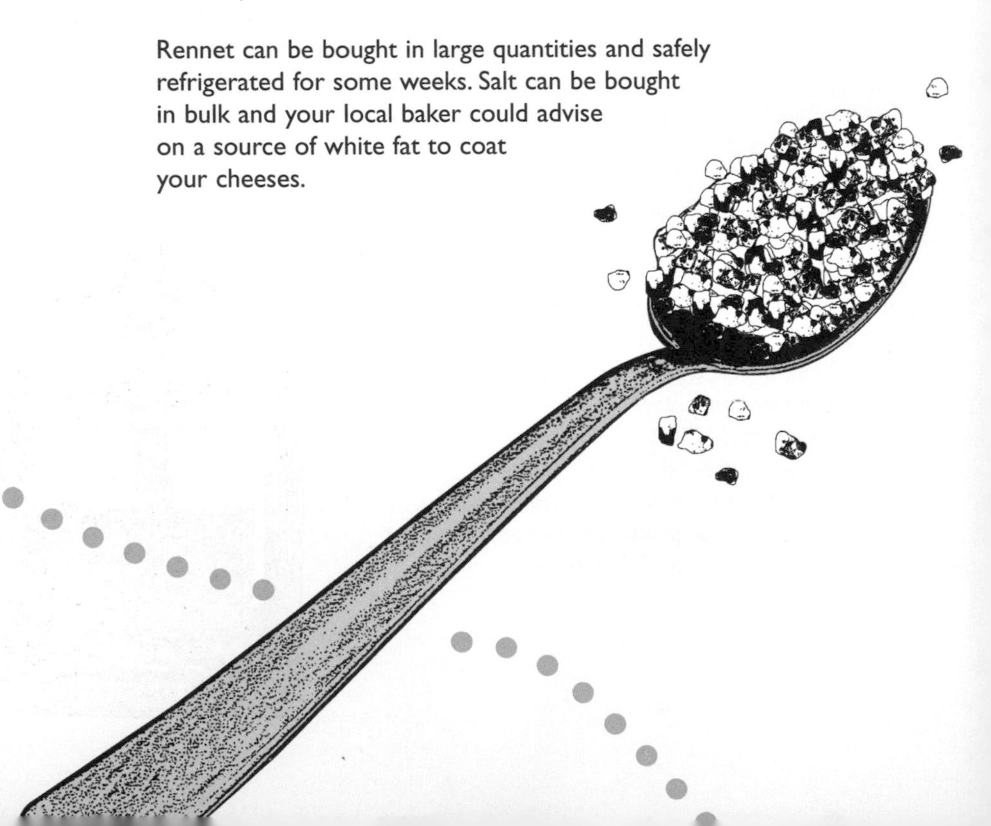

Where are you going to store the cheeses after making?

The cheese storage area should be immediately connected with the making area or part of the same room but separated from it with plastic curtains. This is provided that it does not become too warm during times of production. The cheeses should be stored on slatted shelves or grids to allow for the circulation of air. These shelves should be easy to clean and are often made of planed and smoothed odourless timber (however, you should check the permitted materials with your local environment officer).

What are you going to call your product?

The name you give to your cheese is, of course, your choice, but it is best to avoid classic names such as Cheddar, Cheshire, Stilton, Coulommier etc. as many of these have specific requirements as to composition, texture and even production. If you can, research and make cheese according to an old recipe that has gone out of production as this would give it a greater prestige. Make certain that no other product on the market bears the name you choose and that is has customer appeal. To call your cheese by your own name or the name of your premises is acceptable and sensible and cannot lead to argument or copying. If you are making more than one product, your own dominant name on each e.g. Bloggs Mature Hard and Bloggs Creamy Soft will identify your cheeses from each other.

How will you package, market and distribute your goods?

Packaging

Presentation is important so package your cheese carefully. Soft cheeses need eye-catching containers to make them stand out among the usual tubs. Plain and standard printed pots are available in small quantities from dairy suppliers, but if your production and aspiration can justify it, custom printed packaging is the best. Normally this is only available in large quantities from packaging specialists.

Soft molded cheeses need boxes or a firm supporting band round them before wrapping, whilst Feta and Halloumi are best vacuum packed. You might find a packer who will vacuum pack for you or you could look, with extreme caution, for a second hand machine. The best way to ensure that you have a working machine and one which will 'gas' the packets, if need be, is to hire one from a packaging machine rental firm. Hard cheese will need different sorts of presentation. Whole cheeses should be packaged in rind for cutting to order and portions should be vacuum packed for display selling.

Labelling

The labels on your cheeses are very important. They must impact on the customer and be tempting. However, labelling is seriously regulated and what has to be stated and cannot be implied is a matter for advice from your local trading standards officer, who will either provide you with or tell you where to find all the relevant rules and regulations. You may be obliged by your local officer to have your cheese tested from time to time for composition and bacterial count, which is all part of conforming to the labelling laws. It is therefore worthwhile to locate your nearest testing laboratory, which may not be run by your local authority, and to factor in the cost of testing when pricing your cheese.

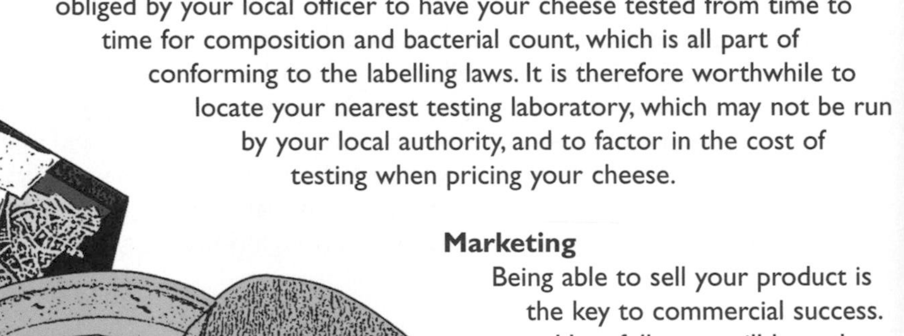

Marketing

Being able to sell your product is the key to commercial success. Hopefully, you will have the beginnings of your market among family and friends before you start to expand. The local markets and farmer's markets are good outlets, but remember that the latter needs other sellers as well as buyers to support them. A private, local round is useful and some privately-owned delicatessens are sometimes willing to stock a new line. In a holiday area, a small cheese is a welcome and useful take-home gift or some restaurants may be interested in adding it to their cheese boards.

Distribution

When you have filled all of these outlets you will need to contact a cheese factor, whose business it is to know the trade and sell your cheese. A factor will work on a percentage payment but this can soon be less expensive than your own time away from production. Be prepared to offer samples of cheese for tasting but be wary of requests from retailers for whole cheese samples who really have no intention of buying. Unfortunately, soft cheeses, which are the least expensive to produce, are the most difficult to distribute for the beginner on anything but a local basis. They need refrigeration both in store, during transport and whilst on display and are subject to spoilage and damage during their short shelf life.

Where and to whom will you sell your cheese?

On the whole, resist the temptation to canvass large and prestigious retailers. The response is likely to be poor, and should it be successful, the volume they demand could very well outstrip your capacity. In addition the company's requirements and transport costs can radically alter your costings. By all means distribute promotional material, but a number of small specialist producers are known to have entered into contracts involving expensive expansion, only to find the contracts not renewed and a surplus of product on their hands.

But if your product is good and you are reliable, there is money to be made in cheese-making as well as much pride and pleasure to be experienced.

Good luck!

Glossary

ACID Lactic acid in milk and curd is produced by the action of bacteria on lactose (see Lactose on page 119).

ACID CURD A simple soft cheese made from soured milk with little, if any, rennet.

ACIDITY The measurement of acid in milk or curd. This used to be done by chemical titration but is now more usually measured with a pH meter or indicator papers.

ALBUMIN Milk contains two main proteins, casein (see opposite) and albumin. Casein coagulates in the presence of acid and the enzyme rennin, but albumin is resistant to these acids to coagulation and remains in the whey unless it is exposed to high temperatures.

ANNATO Annatto is a yellow/orange/red vegetable dye used to colour cheese. A small amount is sometimes added to winter milk to give a golden colour to the cheese, and larger amounts are used in Red Leicester and Gloucester.

BLOCKS (of curd) The mat of curd is cut into squares or oblongs known as blocks after it has settled into the bottom of the vat and the whey has been removed. The size of the blocks required will be given in the recipe. Block also refers to manufactured cheese pressed into rindless rectangular blocks for cutting into portions.

BLUE Blue refers to the Penicillin Roquefort mould culture added to the curd to produce cheeses such as Stilton, Dorset and Roquefort. Blue veins are the lines of blue mould growth, ideally radiating throughout the cheese after maturing. Blue mould also grows on the surface of cheeses in store and can be cleaned off leaving a grey/blue rind.

BREAKING This is a term used when blocks of curd are separated in the vat after they have spread and stuck together during draining. This is sometimes done with a knife (cut) but more often by hand (broken). Curd is also broken or milled into small pieces before salting, moulding and packing into molds.

BRINE Some cheeses are immersed in brine. Brine is a solution of pure salt in clean water, usually a 20 per cent solution made by weighing 200 g (7 oz) of salt, dissolving it in less than 1 litre of hot water and making the solution up to 1 litre (1¾ pints) with cold water.

CASEIN From the cheese maker's point of view, the most important protein in milk which will coagulate in the presence of acid and/or rennin and from which the majority of cheese curd is produced.

CHEDDARING The process which derives from the Cheddar cheese making method of blocking and turning the blocks of curd in the drained vat until all but the last of the whey is drained.

CHEESECLOTH A closely woven soft cotton cloth which is highly absorbent and is used for straining, draining, lining molds and removing the last of the surplus whey from the vat. Disposable cheesecloth of man-made fibre is also available for use in lining molds but is not as good for other purposes. Disposable or not, it can be washed and is reusable.

COAGULATION The action of acid and/or an enzyme, e.g. rennin, on milk, which causes it to take on a semi-solid form.

COAT The outer skin of a pressed cheese or the mould growth on a surface moulded cheese.

CREAM The butterfat and solids portion of the milk which can be removed by hand-skimming from the surface of the milk after standing for several hours. It can be separated from un-stood milk by centrifugal action (spinning) if passed through a separator.

CURD The name for coagulated milk after the action of acid and/or enzyme and also for the remaining solids when the whey has been removed.

CUTTING Cutting the curd after coagulation is the manner in which the curd is reduced to cubes of various sizes. It is easiest when done with curd knives; either a pair of one vertical and one horizontal cut or a single diagonally bladed knife. It can be achieved with a conventional single bladed knife followed by a wire bent at right angles, but this is time consuming.

DAMAGE (to cheeses) The coats of cheeses, especially when young, can be easily damaged by mis-handling. A knock is often enough to break the coat or rind and so allow moulds to penetrate and grow in the cheese. All damage should be repaired by filling the marks with solid fat or butter, or packing with dry salt.

DRAWING THE WHEY This action refers to the process of separating the whey from the draining curd. Larger vats will have a whey drain, but small vats and other vessels without a whey drain need to be baled out with a jug or bowl and the dregs soaked up with cheesecloth.

ENZYME The enzyme rennin is extracted from the mucous membrane of the stomach of calves, lambs and other suckling animals. It is used to prepare the animal rennet for cheese making. Non-animal rennet is sourced from a microbial enzyme with similar coagulation properties.

FAT Butterfat is held in suspension in milk in the form of minute globules. It is a complex mixture of fats and is a liquid over a wide range of temperatures. It rises readily to the surface of the milk and can be easily wasted in the whey unless handled carefully. This is the reason for 'top stirring'. No such problems occur with homogenized milk in which the butterfat is held captive. Fat for greasing the rinds of pressed cheeses can be butter, lard or solid vegetable oil or replaced with good quality olive oil.

FOLLOWER This is the name given to the plate or disc which is placed on the top of the cheese in the mold which bears the weight from the press and spreads the pressure evenly throughout the cheese. First pressings will have the follower outside the folded over cheesecloth, final pressings have the follower directly on the cheese. It is important that followers are rigid under pressure and are a sliding fit inside their molds.

HEAT TREATED While the regulations vary from country to country, it is usual that milk can be called pasteurized only if an automated record of its correct treatment is available for inspection. However, milk heated to 68° C (154° F) and held there for 30 minutes or 72° C (162° F) and held for 30 seconds before use or cooling, without a record of the process being recorded, can be referred to as 'heat treated'.

HOMOGENIZED Milk which has been mechanically treated to mix the fat with the rest of the constituents and to prevent its separation later.

LACTOBACILLUS The micro-organisms present naturally in milk before pasteurization or heat treatment. They are, of course, present in large numbers in starter.

LACTIC ACID A by-product of the action of lactobacillus on lactose or milk sugar.

LACTOSE The natural sugar present in milk from which lactic acid is produced by the action of lacto bacillus. The milk is then referred to as 'soured' or 'ripened'.

MICRO-ORGANISM Bacteria, moulds and viruses together with some rarer organisms are all referred to loosely under this heading. Those which cause illness are called pathogens.

MICROBIAL Starters for cheese (see Starter Cultures on page 122) are a selected mix of micro-organisms, prepared in the laboratory and freeze-dried for keeping. They replaced the liquid mother cultures of the past. They are sometimes referred to as microbial cultures.

MILLING Refers to the breaking or chopping of the drained curd into small pieces before packing into molds. Milling can be done by hand or machine, the latter either hand operated or electrically driven.

MOLD In this book the U.S. spelling has been used for the formers into which the milled curd is placed to press, shape and exclude the last of the whey from the worked curd. It is also used to refer to the formers, tubes and baskets used for soft cheeses. This is to avoid confusion with Mould below.

MOULD Moulds are micro-organisms which are slightly larger than bacteria. They give rise to various growths in and on cheese. Those which appear as dry blue, green, grey or white are common and harmless on the rinds of hard cheese and can be readily cleaned off with a dry cloth or paper towel. Wet, sticky, highly coloured or smelly moulds should be vigorously controlled by washing, drying and salting the rinds of the affected cheeses. Blue mould (see Blue on page 116) is used to produce blue veined cheeses and white mould, which grows like felt on the surface of soft cheeses, is used when producing Camembert or Brie.

MYSOST This is whey cheese produced from boiling the whey residues from cows' milk cheese.

PASTEURIZATION This term refers to the reduction of the bacteria in fresh milk to a safe level by submitting the milk to heating over a minimum time. A high temperature/short time is favoured as less damage to flavour occurs. Pasteurization is not sterilization.

PATHOGENS Those bacteria which cause illness in human beings.

pH A measure of the degree of acidity. In the case of cheese making, the measurement is taken in the whey. The scale has 7 as neutral, with values for acidity falling and alkalinity rising. The measurement is taken with a pH meter, indicator papers (available from sundries suppliers) or in a factory with an in-line recorder. pH values are not mentioned in this book where a sense of smell and judgement are called for.

PITCHING Part of the cheese making process when the curd is allowed to sink to the bottom of the vat after cutting where it forms a mat with the whey above.

PRESSURE Applying weight/pressure to the packed curd in the molds to produce a shaped and homogeneous cheese.

PROTEINS There are two major proteins in milk considered under the headings of Albumin (see page 116) and CaseIn (see page 117).

RENNET The refined enzyme rennin.

RICOTTA Ricotta is the precipitated albumin from whey, sometimes with added milk, brought about by near-boiling temperatures until the Ricotta floats to the top of the whey. It is then skimmed off, pressed and dried.

RIND The term defining the hardened coat of a finished, and usually mature, hard cheese.

RIPEN The process of ageing and maturing a cheese in store which leads to improved and defining flavours. Also used to describe the process of souring or acid formation after starter is added to the warm milk in the vat.

SALT Salt is used for flavouring and, to a lesser degree, preservation. It can be added to the curd before packing into the mold, rubbed on the surface of the cheeses after molding or pressing, or made into brine into which the cheese is immersed. The choice of salt is wide, but since block salt disappeared from the shelves, it seems that every variety has a free running and anti-caking additive which some say imparts a bitter flavour to the cheese. For this reason, table salt should be avoided and cooking salt preferred as it usually contains less additives. Sea salt attracts moisture to the coat and is best avoided.

SET/SETTING Setting or coagulation follows the addition of rennet to milk as in the preparation of junket. It also occurs without rennet when a sufficiently high acidity is reached through the action of lactic bacteria. A combination of both acid and rennet produces the firm curd desirable in cheese making.

SOUR/SOURING The conversion of the lactose in 'sweet' milk to the lactic acid in 'sour' milk.

STARTER The selected microbial mix specially formulated and prepared for cheese making. The starter used for the cheeses described in this book are all prepared before use by adding freeze-dried bacteria and medium to boiled and cooled whole milk, the whole then kept warm overnight until the mixture has set. DVI or direct vat inoculation is an alternative whereby the specially prepared bacteria and medium are added in the dry state to the milk in the vat.

STILTON KNOT This is a description of the method of taking three corners of a cloth in which curd has been bagged, gathering them together and looping and tying the forth corner round the other three so that the knot can be tightened at will.

STIRRING Stirring both the milk and curd is best done gently even when thorough stirring is necessary. A special rake can be used, but on a manageable scale it is better done by hand when the 'feel' is important.

TOP-STIR This is the action of keeping the top surface of the milk moving with the tips of the fingers to prevent the cream rising to the surface and so being excluded from the forming of the curd. It is unnecessary with homogenized milk.

VAT OR VESSEL The container, once wooden, copper or tin plate and now usually made of stainless steel, in which the cheese is made. Ideally this should be water jacketed with controlled heating, but cheese can be made in the most primitive of vats. A whey drain which exits through the water jacket is a must for large quantity production and a square or rectangular shape, not deep in proportion, is an advantage when cutting and blocking the curd. Beware of bulk milk tanks which have been modified with too great a depth for the hands to reach the bottom.

WASHING All equipment used in cheese making must be thoroughly cleaned before use. Small equipment which comes into contact with the curd, such as knives, should be plunged into cold water immediately after use to prevent the curd residues hardening. Rinsed free of curd particles any equipment which is small enough is best passed through a dishwasher on a full hot program using top quality tablets or powder. Larger equipment washed by hand using a good detergent and paying particular attention to corners and crevices, should be rinsed well in clean water and allowed to drain dry. Many cheese makers use a final rinse of weak bleach (hypochlorite) to sterilize but, theoretically at least, this should be rinsed again with clean water. Cotton cheese cloths need rinsing to remove adhering curd from the weave and can then be put through a washing machine or on a hot wash.

WATER All the water used in cheese making should be of potable quality, particularly that added to the cheese when diluting the rennet or as 'washed curd' water.

WHEY The liquid which is produced when the milk is coagulated and the curd is cut. Whey escapes from every surface of the cut or ladled curd and is run or baled off and either discarded or utilised in some way. Whey should not be put into domestic drainage systems including septic tanks but can be fed to some livestock and pets, added to the compost heap or spread on the land.

WHITE MOULD The mould used in Camembert and Brie production and in other cheeses with a white mould coat. It is usually Penicilium Camemberti or Penicillium Candidum, which are surface growing moulds that have a considerable effect on the internal ripening and creamy texture of the cheeses.

WAX Some cheeses are coated with highly coloured wax. Edam made for export is traditionally red wax coated and many small producers use black, green, yellow or orange wax, all available from cheese making suppliers. Coating cheese successfully without it developing mould beneath the wax is sometimes difficult and should not be attempted without advice and trials.

YIELD Yield is the amount of finished cheese produced from any given quantity of milk. The yield varies from variety to variety of cheese and sometimes within the same variety depending on the milk and the variations in the processes.

Suppliers

Ascott Smallholding Supplies Ltd
Starter, rennet, presses, molds etc. A good range of equipment and sundries.
www.ascott.biz
Tel: +44 (0)845 130 6285 Fax: +44 (0)870 774 0140
The Old Creamery, Four Crosses, Llanymynech, Powys, SY22 6LP

Ash, AG & RH
Curd mills and curd knives. Advice on sourcing equipment, consultancy and advice on cheese-making.
Tel: +44 (0)1404 881529
Millhayes Mill, Stockland, Honiton, Devon, EX14 9DB

Jongia (UK) Ltd
A range of cheese-making equipment and supplies.
www.jongiauk.com
Tel: +44 (0)121 7444844 Fax: +44 (0)871 7503626
23 Prospect Lane, Solihull, Birmingham, West Midlands, B91 9DB

Nisbets plc Catering Equipment
Extensive catalogue including thermometers, bains marie and small tools.
www.nisbets.co.uk
Tel: +44 (0)845 1405555 Fax: +44 (0)845 1435555
Fourth Way, Bristol, Avon, BS11 8TB

Strattons Sales Ltd
Cheesecloth, starter cultures, rennet and wax.
www.strattonsales.co.uk
Tel: +44 (0)1749 838690
8 Leighton Lane Industrial Estate, Evercreech, Somerset, BA4 6LQ

Wheeler R & J (Engineers)
Wheeler press and vats made to order.
Tel: +44 (0)1392 832238
Hoppins, Dunchideock, Exeter, Devon, EX2 9UL

Acknowledgements

This book is written with grateful thanks to all those people who have encouraged me to practice my craft for the last thirty years. To my students who asked the questions to which I had to find the answers, and to those experienced cheese makers and authors who helped to supply the answers.

My gratitude must be extended to my husband for his practical skills when making equipment to solve problems and to Catherine Rands for typing and putting the whole text into computer shape – a skill completely beyond me!

I am truly grateful to the staff of The Magdalen Project for their interest and support and their facility for cheese-making courses. Thanks, too, to Tabb House publishers for allowing the publication of another cheese-making book after the original *Cheesecraft*.